Mathematics

A PRACTICAL GUIDE TEACHER BOOK

Talei Kunkel
Palmira Mariz Seiler
Marlene Torres-Skoumal
David Weber

International Baccalaureate
Baccalauréat International
Bachillerato Internacional

Mathematics: A practical guide (Teacher book)

Published on behalf of the International Baccalaureate Organization, a not-for-profit educational foundation of 15 Route des Morillons, 1218 Le Grand-Saconnex, Geneva, Switzerland by the International Baccalaureate Organization (UK) Ltd, Peterson House, Malthouse Avenue, Cardiff Gate, Cardiff, Wales CF23 8GL United Kingdom, represented by IB Publishing Ltd, Churchillplein 6, The Hague, 2517JW The Netherlands. Website: www.ibo.org

The International Baccalaureate Organization (known as the IB) offers four high-quality and challenging educational programmes for a worldwide community of schools, aiming to create a better, more peaceful world. This publication is one of a range of materials produced to support these programmes.

IB merchandise and publications can be purchased through the IB store at http://store.ibo.org. General ordering queries should be directed to the Sales and Marketing Department at sales@ibo.org.

British Library Cataloguing in Publication Data

A catalogue record for this book is available from the British Library

ISBN: 978-0-9927035-1-6
MYP338
Typeset by Q2A Media Services Pvt. Ltd.
Printed in India.

Impression 4
Year 2016

Acknowledgments

To my husband, James, for his support and understanding throughout the long writing process. And to my daughter, Kate, for still smiling even though I missed many hours of fun and playing games. When it is your time to go to school, I hope textbooks such as this will spark your interest and enjoyment in learning mathematics and solving mathematical problems. — Talei

To my family and my husband in particular for all the support and love. Thanks for feeding my creativity, sharing my enthusiasm and coping with my commitments to this project throughout our summer break. — Palmira

Conceiving new tasks for such an ambitious project involved many weekends of work during the nicest months of the year, so many thanks to my husband Peter for his endless patience and stimulating conversations in applying the beautiful and perfect world of mathematics to the less than perfect real world. — Marlene

Thanks to my family, friends, colleagues and students for their patience, support and inspiration in everything I do. — Dave

We are grateful for permission to reprint copyright material and other content:

p60 hut: http://commons.wikimedia.org/wiki/File:Palawan_-_Tropical_Hut.jpg; p63 Mt. Vesuvius: http://commons.wikimedia.org/wiki/File:Vesuvius_from_Naples_at_sunset.jpg; p65 blood transfusion: http://commons.wikimedia.org/wiki/File:US_Navy_060505-N-2832L-056_Lt._Angela_Banks_draws_blood_from_a_mannequin_during_training_for_antilogous_blood_transfusion_with_a_cell_saver_in_one_of_the_12_operating_rooms_aboard_Military_Sealift_Command_(MSC)_hospital_ship_U.jpg; p78 clinometer: ©David Weber; p95 Pascal's triangle: http://commons.wikimedia.org/wiki/File:Bw3i.jpg; p101 galaxy: http://commons.wikimedia.org/wiki/File:Hubble2005-01-barred-spiral-galaxy-NGC1300.jpg; p118 Mathematikum: HYPERLINK "http://commons.wikimedia.org/wiki/Category:Mathematikum" \l "mediaviewer/File:Giessen_Liebigstrasse_8_61537.png"http://commons.wikimedia.org/wiki/Category:Mathematikum#mediaviewer/File:Giessen_Liebigstrasse_8_61537.png; p122 rock, paper, scissors: http://commons.wikimedia.org/wiki/File:Rock-paper-scissors.svg; p132 Leaning Tower of Pisa: ©iStockphoto; p135 Petronas Towers: ©iStockphoto; p137 Earth: http://commons.wikimedia.org/wiki/Category:Earth_from_space#mediaviewer/File:Apollo_13_-_Earth_during_journey_home.jpg; p140 group work: ©Andreser shutterstock.com Image ID: 106222658; p145 *Treatise on Mensuration*, Albrecht Dürer, 1538; p149 Geodesic dome: ©iStockphoto; p161 boat: ©iStockphoto.

IB learner profile

The aim of all IB programmes is to develop internationally minded people who, recognizing their common humanity and shared guardianship of the planet, help to create a better and more peaceful world.

As IB learners we strive to be:

INQUIRERS
We nurture our curiosity, developing skills for inquiry and research. We know how to learn independently and with others. We learn with enthusiasm and sustain our love of learning throughout life.

KNOWLEDGEABLE
We develop and use conceptual understanding, exploring knowledge across a range of disciplines. We engage with issues and ideas that have local and global significance.

THINKERS
We use critical and creative thinking skills to analyse and take responsible action on complex problems. We exercise initiative in making reasoned, ethical decisions.

COMMUNICATORS
We express ourselves confidently and creatively in more than one language and in many ways. We collaborate effectively, listening carefully to the perspectives of other individuals and groups.

PRINCIPLED
We act with integrity and honesty, with a strong sense of fairness and justice, and with respect for the dignity and rights of people everywhere. We take responsibility for our actions and their consequences.

OPEN-MINDED
We critically appreciate our own cultures and personal histories, as well as the values and traditions of others. We seek and evaluate a range of points of view, and we are willing to grow from the experience.

CARING
We show empathy, compassion and respect. We have a commitment to service, and we act to make a positive difference in the lives of others and in the world around us.

RISK-TAKERS
We approach uncertainty with forethought and determination; we work independently and cooperatively to explore new ideas and innovative strategies. We are resourceful and resilient in the face of challenges and change.

BALANCED
We understand the importance of balancing different aspects of our lives—intellectual, physical, and emotional—to achieve well-being for ourselves and others. We recognize our interdependence with other people and with the world in which we live.

REFLECTIVE
We thoughtfully consider the world and our own ideas and experience. We work to understand our strengths and weaknesses in order to support our learning and personal development.

The IB learner profile represents 10 attributes valued by IB World Schools. We believe these attributes, and others like them, can help individuals and groups become responsible members of local, national and global communities.

International Baccalaureate®
Baccalauréat International®
Bachillerato Internacional®

Contents

How to use this book   vi

1. Introduction to IB Skills   1

2. Introducing key concept 1: Form   10

3. Introducing key concept 2: Logic   13

4. Introducing key concept 3: Relationships   18

5. Change   21

    Topic 1: Finite patterns of change   22

    Topic 2: Infinite patterns of change   25

    Topic 3: Financial mathematics   27

6. Equivalence   31

    Topic 1: Equivalence, equality and congruence   32

    Topic 2: Equivalent expressions and equivalent equations   34

    Topic 3: Equivalent methods and forms   36

7. Generalization   42

    Topic 1: Number investigations   44

    Topic 2: Diagrams, terminology and notation   47

    Topic 3: Transformations and graphs   53

8. Justification   56

    Topic 1: Formal justifications in mathematics   57

    Topic 2: Empirical justifications in mathematics   63

    Topic 3: Empirical justifications using algebraic methods   66

9. Measurement   68

    Topic 1: Making measurements   69

    Topic 2: Related measurements   72

    Topic 3: Using measurements to determine inaccessible measures   75

10. Model   80

    Topic 1: Linear and quadratic functions   81

    Topic 2: Regression models   87

    Topic 3: Scale models   89

11. Pattern   94

    Topic 1: Famous number patterns   96

    Topic 2: Algebraic patterns   101

    Topic 3: Applying number patterns   103

12. Quantity   108

          Topic 1: Volumes, areas and perimeters   109

          Topic 2: Trigonometry   112

          Topic 3: Age problems   114

13. Representation   116

          Topic 1: Points, lines and parabolas   117

          Topic 2: Probability trees   122

          Topic 3: Misrepresentation   124

14. Simplification   127

          Topic 1: Simplifying algebraic and numeric expressions   128

          Topic 2: Simplifying through formulas   133

          Topic 3: Simplifying a problem   136

15. Space   140

          Topic 1: Special points and lines in 2D shapes   141

          Topic 2: Mathematics and art   143

          Topic 3: Volumes and surface area of 3D shapes   147

16. System   150

          Topic 1: The real number system   151

          Topic 2: Geometric systems   152

          Topic 3: Probability systems   156

How to use this book

The Teacher Book is designed both as a companion to the student book and to facilitate a whole school approach to mathematical skill development in Middle Years Programme students. As well as providing definitions and explanations for key concepts, this book includes supporting activities, task guidelines and assessment criteria that have been specified for the tasks.

The student and teacher books provide a detailed introduction to the key and related concepts in MYP mathematics.

The key concept chapters look at the challenges and benefits of teaching for conceptual learning and introduce mathematical skills. Ideas for teaching these concepts and skills are provided, some of these are linked to the student book and form extensions of activities located therein, others are new activities that you can use in your classes to motivate and engage students' understanding of conceptual learning and skills development.

The related concept chapters support the delivery of the related concepts in the classroom. Teacher guidance is provided to complement the activities in the student book.

Throughout the book you will find features and teaching suggestions that will help you link your teaching to the core elements of the MYP. Here are some of the features you will come across:

DIPLOMA PROGRAMME LINKS
Opportunities to link to the DP curriculum.

INTERDISCIPLINARY LINKS
These boxes provide links to other subject groups.

CHAPTER LINKS
MYP students are encouraged to use skills and knowledge from different subject areas. These boxes link to other chapters, which relate to a topic or theme.

 WEB LINKS

The student and teacher books have integrated references to internet tools and sources in each chapter.

TEACHING IDEA

These boxes give additional ideas to the activities in the student book.

QUICK THINK

These boxes refer to the Quick Think in the student book and give further guidance on how to use these suggestions to extend student learning or to facilitate a discussion.

TIP

Throughout the chapters you will see additional tips for teaching.

🏛 MATHS THROUGH HISTORY

These boxes give historical context to some of the key ideas and concepts explored.

1 Introduction to IB Skills

Welcome to the *IB Skills: Mathematics* book for MYP 4/5. This is one book of a set, along with the student book of the same name, that aims to help IB students develop and apply mathematical skills through the use of inquiry-based activities and authentic, real-life tasks. Rather than being aimed at one particular course, this book includes a wide range of activities from all four branches of the Middle Years Programme (MYP) mathematics framework: number, algebra, geometry and trigonometry, and statistics and probability. Many of the activities in the teacher and student books are aimed at a typical "standard level" mathematics course, while others may seem to be at a more "extended level". This variety is intended to allow you to pick and choose those activities that best serve the needs of your students, while also preparing them for future mathematics courses, including the Diploma Programme. Of course, activities may be adapted for students of different abilities or even different grade levels.

Whereas the student book contains instructions for students, questions for reflection and opportunities for research, this teacher book is intended to explain how to use those activities while providing optional tasks, information for unit planning and extensions, among other suggestions. The IB Skills book is a valuable resource, not only for teaching and learning in MYP mathematics, but in any classroom.

Structure of the books

Both the student book and this teacher book are organized by related concept. Each of the 12 related concepts in MYP mathematics is explored in its own chapter, through a wide range of activities. The introductory table at the beginning of each chapter in this teacher book presents the topics covered and a skills checklist. In the student book, a similar table is accompanied by the relevant command terms and other related concepts. At a glance, teachers can see which topics are explored and the mathematics learned and/or applied, making it easier to decide which activities they might want to use.

Every chapter continues with a quote, generally from a mathematician, related to the concept being developed. Chapters are then subdivided into topics that help students discover a different facet of the related concept. At the end of most activities or topics is a "Reflection", giving students the opportunity to think about what they have learned and how it may relate to what they already know.

It is not necessary for students to do every activity in every chapter in order to understand the related concept. Generally, activities are independent of each other and can be used as necessary, depending on the unit being taught. However, this teacher book should be seen as a companion to the student book. The general structure is the same as that of the student book, although it also gives advice on how to use the student activities, as well as other optional activities.

Where applicable we have included activities that could be used as a summative task for a unit. For these activities, a portion of Stage 1 of the unit planner is provided in this book. Since all of these tasks are real-life problems, they could be assessed with Criterion D. If the task is submitted as a report (for example) that requires logical structure and allows multiple forms of representation to present information, then Criterion C could also be used to assess the task.

Key concept	Related concepts	Global context
	Statement of inquiry	

Unit planning in mathematics

While MYP mathematics has a framework of suggested topics, there is no set curriculum. It is more of a philosophy of teaching that can be applied to any curriculum, whether chosen by the school or prescribed by a local or national agency. However, MYP teachers in all subject areas are expected to design units of study that not only address the selected curriculum, but also incorporate the philosophy of the MYP. This section gives suggestions on the different elements of the unit planning process.

The conceptual side

When you decide on content for a unit, it has to be set within the conceptual framework. To do that, a key concept is selected which, in mathematics, is form, relationships or logic (although any of the 16 MYP key concepts may be used). One or more pertinent, related concepts are also chosen from the 12 that are developed in this book. The choices here are entirely personal because it is based on the teacher's philosophy of how they view the learning of the content. One teacher may classify solving equations as "logic", while another views it as "relationships", while yet another may see it as "form". The "conceptual side" is grounded in the choice of key and related concepts.

In chapter 1: "Introduction to IB skills" in the student book, students are introduced to this conceptual framework with an activity in which they try to select the key and related concepts for units they studied in the past. Typical answers are given below, although they should not be considered as correct answers. Once again, it is dependent on the philosophy and choice of the teacher.

Unit	Key concept	Related concepts
Solving one- and two-step equations	Logic	Simplification Equivalence
Decimals, ratio and percentages	Form	Equivalence
Perimeter and area of two-dimensional shapes	Relationships	Measurement
Order of operations	Logic	Quantity Simplification Equivalence
Mean, median and mode	Relationships	Quantity Generalization Justification

The contextual side

The focus of the MYP has always been to teach in a context, which is never "because the test is on Friday". Contextualized learning is developmentally appropriate for 11–16-year-olds and aligns very well with what students experience in the Primary Years Programme. There are six global contexts, each with its own set of descriptors. In mathematics, the selection of the appropriate global context is often made easier by setting an appropriate summative task first. It may be entirely possible to select a topic, such as linear functions, and then a global context, such as personal and cultural expression, and to design a summative task that explores the chosen context in sufficient depth. However, it is often easier to select the content (linear functions) and then outline the task, such as predicting the eruptions of a volcano, before choosing an appropriate global context, such as scientific and technical innovation. In fact, this is an effective indicator to check whether or not the proposed summative task is authentic enough. If it is easily described by one of the global contexts, then it is probably an appropriate task.

The statement of inquiry

The key concept, the related concepts and the global context are all summed up in a "statement of inquiry". This is one sentence that has both the mathematical concepts and the context in it. (This is probably the most difficult part to do.)

Key concept + at least one related concept + global context = statement of inquiry.

When writing the statement of inquiry, you could use one of the six global context titles, one of its descriptors or even one of the explorations. For example, you may have selected identities and relationships as your global context, but decide to use beliefs and values (one of its descriptors) in the statement of inquiry. Adaptation is also possible: for example, one teacher may choose the portion that states "what it means to be human" while another may adapt it to something more specific, such as "what it means to be a minority". While this last phrase isn't specifically stated in the descriptors, it definitely fits and can be used in the statement of inquiry. The global contexts are, in fact, incredibly flexible.

Relating the summative task to the statement of inquiry

In the unit planner, there is a specific section in which you have to explain and reflect on how the summative task(s) relates to the statement of inquiry. Selecting the task(s) first, as described above, guarantees a relationship between the task(s) and the statement of inquiry. Ultimately, each unit should have a global context that is explored throughout the unit, including in a summative task in which students apply the mathematics that they have learned to an authentic situation.

Inquiry questions

You are also expected to develop a total of three to five inquiry questions for each unit, on three different levels: factual, conceptual and debatable. The factual questions generally are derived directly from content taught in the unit and they have an answer that students can either recall or look up. (For example: How are the slopes of parallel lines related? What are the necessary conditions to prove two triangles congruent?) Debatable questions are more related to the global context of the unit. As students learn content and explore the global context, they begin to be able to answer a more provocative question that doesn't focus as much on the mathematics. (For example: What does it mean to be "fair"? What is more natural, order or chaos?) The conceptual questions lie between the two of these, going beyond mere facts but remaining more mathematical and less provocative. (For example: How are the trigonometric ratios related to similar triangles? How do the different systems of numbers relate to one another?) These inquiry questions guide both the teacher and the students through the unit.

Assessment criteria

All subjects have four criteria, each marked out of 8. The criteria match the objectives for the subject, which are the following in mathematics.

Criterion A: Knowing and understanding

It is appropriate to use this criterion with tests and problems that are not authentic, real-life applications. In order for students to achieve at the highest level, there must be some unfamiliar problems for them to solve.

Criterion B: Investigating patterns

It is appropriate to use this criterion with investigations (which could be purely mathematical or involve a real-life application) in which students find a general rule and then prove or justify it. In order for them to achieve at the highest level, students must select and apply their own problem-solving techniques and prove or justify the rule that they develop. Tasks that are too guided or scaffolded do not allow students to reach the highest level.

Criterion C: Communicating

It is appropriate to use this criterion with investigations and real-life problems in which students have an opportunity to use different forms of mathematical representation (graphs, equations, tables, and so on) in an attempt to communicate their mathematical operations and thinking. This criterion also applies when students submit a report, for example, since the work has to have a "logical structure". Well-designed tests may also be assessed against criterion C, provided that students can reach the highest descriptors of the criterion.

Criterion D: Applying mathematics in real-life contexts

It is appropriate to use this criterion with any real-life problems or tasks. Once again, students must select and apply their own problem-solving techniques. If the task is an authentic, real-life application of mathematics, then it is likely to be assessed with criterion D. Modelling tasks may be assessed with this criterion as long as the data is authentic.

Teachers are free to combine criteria in assessing student work. Criterion C can be paired with criteria A, B or D, as explained previously. Generally, criteria B and D are not assessed together, unless one part of the task requires students to discover a general rule that they then apply to a real-life situation. (This could also be assessed as two separate tasks.) Task-specific clarifications should be given, not only to make applying the criteria easier, but so that students know exactly what is expected of them. A teacher who has a test and an authentic assessment task, and who asks students to investigate some of the content in the unit, can easily use all of the assessment criteria in any unit (though this is not a requirement).

Teaching and learning in the MYP

Teachers in MYP mathematics classes are encouraged to use inquiry to help students discover content and concepts, develop techniques and acquire skills, rather than simply having information disseminated to them. You are expected to use a wide variety of teaching techniques (including lecture where appropriate), but the focus is still on the student as the learner, not on the teacher as the sole provider of knowledge. Therefore, most of the activities in the student book are inquiry-based, with even more inquiry lessons suggested in this teacher book. However, there are also activities that are more "theoretical" in nature, allowing students to make generalizations and develop abstract reasoning that are prerequisite skills for the Diploma Programme.

Summary of activities in the student book

All of the activities in the student book are categorized below, using the four branches of the MYP mathematics framework. Some have been listed in two branches because those activities include tasks that encompass mathematical content from both.

Chapter	Branch: Number
Change	Discover arithmetic patterns of change and use formulas to add consecutive terms of arithmetic sequences. Discover geometric patterns of change and how rapidly geometric sequences increase or decrease. Explore different investment options (arithmetic and geometric sequences) and factors that must be taken into account when investing money.
Equivalence	Use exchange rates to calculate values between currencies and look at the Big Mac index as a means to compare countries.
Generalization	Discover numerical patterns in given contexts. Discover a numerical pattern associated with a bad apple spoiling the rest of the tray, given its position and the configuration of the tray. Practise caution in making generalizations, and analyse steps in reasoning to find fallacies in making generalizations in some simple numerical patterns and statements.
Measurement	Approximate the volume of air in a hot-air balloon by using something else in the picture as a reference.
Pattern	Investigate some special numbers and number patterns such as Pascal's triangle, Phi and triangular numbers.
Representation	Use a parabola to represent the real numbers graphically and, in particular, the prime numbers, by using points on the parabola and the y-axis to generate the times table.
Simplification	Use the laws of exponents and logarithms to simplify algebraic and numeric expressions while investigating the geometric mean and performing dimensional analysis.
Space	Finding the percentage increase and decrease of material used after changing the configuration of objects in a package.
System	Investigate the properties of the real number system.

	Branch: Algebra
Change	Apply arithmetic and geometric sequences while investigating different investment options and factors one must take into account when investing money.
Equivalence	Determine the point of intersection of a linear system to compare and analyse regular, hybrid and diesel cars.
	Simplify a system of equations to investigate the correct proportions of different foods to achieve a balanced diet.
	Analyse the different forms of quadratic equations to determine which form is best to use, given the context of the question.
	Model a football kick with a quadratic function to determine if a record has been made.
Generalization	Discover the transformations of rational functions of the form $y = \dfrac{a}{x-h} + k$, their graphs and elements, then make a summary of generalizations about the different transformations and validate these general rules.
Justification	Use axioms and theorems to create a sequence of logical arguments that prove whether a mathematical conjecture is correct, from using simple algebra in direct proofs, to indirect proofs using contradiction and counter example.
	Graph a rational function to make informed decisions about real-world topics (in this case the trade-off between monetary budget and astronaut safety) and justify the choice.
Model	Reproduce a picture made of the graphs of different functions, using either the GDC or software, by defining the appropriate functions and their domain.
	Model a parabolic tunnel, and consider size limitations of trucks passing through the tunnel.
	Evaluate how well a quadratic function models a dental arch and then explore how an "average bite" can be used to make mouth guard moulds.
	Use exponential functions to model both the spread of dengue fever and the results of one method of trying to eradicate the mosquitoes that carry it.
Pattern	Discover some relationships between numerical and algebraic factorization.
	Determine linear, quadratic and exponential patterns in different real-life contexts such as mosaic tile design, the growth of bacteria and the decay of certain compounds and explore how planets may follow a number pattern.
	Play the chaos game to discover underlying patterns in apparently chaotic systems.
Quantity	Determine the optimal design for conical paper drinking cups that will maximize volume.
	Graph the sine function and understand its characteristics.
	Describe the transformations of a sine function in the form, $f(x) = a \sin b(x) + c$.
	Model the periodic motion of a rotating bicycle tyre, using the sine function, and analyse how the tyre size and design of the bike affect its performance.
Representation	Look at equations of lines and further explore their meaning through different representations.
	Explore the differences between Cartesian and vector lines in context by looking at the paths of ships.
Simplification	Solve higher-order polynomial equations by performing long division and factorizing.
	Simplify complex equations, given a specific context such as carbon dating, and calculating how old the Earth is.
	Use a linear programming model to help analyse a situation and make business decisions.
Space	Analyse the Koch snowflake to find relationships between the iteration stage and the length of one side, the perimeter and the area.

System	Create a geometric system that serves to reinforce the key components of a mathematical system: undefined terms, defined terms, axioms and theorems.

	Branch: Geometry and trigonometry
Equivalence	Analyse medial triangles and test for congruency. Analyse congruent triangles in structures.
Generalization	Construct simple and clear diagrams and use appropriate notation to explore networks and analytic geometry problems, and hence discover patterns and rules and communicate general statements associated with them.
Justification	Use axioms and theorems to create a sequence of logical arguments that prove if a mathematical conjecture is correct, using simple geometry in direct proofs. Create visual models to conduct visual proofs and convert them into a formal proof and justify all the steps, using results that are previously proven to be true.
Measurement	Discover theorems in circle geometry, including relationships between tangents to a circle and its radius, lengths of segments of chords, points external to the circle and cyclic quadrilaterals. Use measurements to determine inaccessible measures, such as tall structures and the circumference of the Earth.
Model	Apply trigonometry and the areas of triangles, circles and sectors to a scale model representing a sprinkler system in an attempt to minimize water usage. Apply analytic geometry techniques (triangle centres, shortest distance from a point to a line and so on) to design a residential community.
Pattern	Play the chaos game to discover underlying patterns in apparently chaotic systems.
Quantity	Find the maximum area that can be created by a rectangle with a fixed perimeter. Determine the optimal design for conical paper drinking cups that will maximize volume. Investigate radians as a unit of measure and the relationship to measuring in degrees.
Simplification	Use existing knowledge of right-angled triangles to develop new formulas that can simplify future problems with these shapes, including deriving formulas for metric relations and problems involving multiple right-angled triangles.
Space	Explore and construct steps needed in finding the centroid of different 2D objects (especially triangles). Explore the properties of 3D objects and impossible objects. Investigate the properties of fractals, in particular the Koch snowflake. Geometric sequences and series will be used to illustrate a seeming paradox of an object having infinite perimeter and finite area. Use geometric and trigonometric formulae to solve a real-life optimization problem involving the economics of packaging.
System	Compare axioms, terms and definitions in geometric systems.

	Branch: Statistics and probability
Change	Explore the game of Rolling Dice (Craps) and study the probability of winning this game, given different scenarios.
Justification	Use linear regression and piecewise functions to make predictions about real-world topics (in this case, eruptions of Mount Vesuvius) and justify calculations.
Model	Use linear regression to analyse the winning times in the men's and women's 100 m races in the Olympics. Evaluate how well a quadratic function models a dental arch, using quadratic regression, and then explore how an "average bite" can be used to make mouth-guard moulds. Model both the spread of dengue fever and the results of one method of trying to eradicate the mosquitoes that carry it.
Quantity	Calculate mean, mode and median of student ages. Conduct a graphical analysis of histograms.
Representation	Construct tree diagrams to help solve conditional probability problems that occur in games. Analyse the effectiveness of a medical test by calculating the conditional probability of specific results. Analyse the advantages and disadvantages of the measures of central tendency and the appropriate use of each. Analyse standard deviation and how graphs can be displayed in order to misrepresent data.
System	Beginning with defined terms and axioms, prove basic probability theorems. Given a set of axioms, use basic probability theorems to solve scenarios involving blood type and transfusions.

Teaching mathematics in the Middle Years Programme is not synonymous with the traditional view of opening a textbook, starting at page 1 and continuing until the end of the book. MYP mathematics teachers are designers of activities that enable students to experience the joy of discovery and success in mathematics and then apply it to the world around them. Students understand the mathematics they are learning because they have, mainly, formulated the concepts themselves rather than just hearing about them. Students are more willing to persist in problem-solving because they are so used to meeting new content or new contexts that challenge them to work through any difficulties. At the same time, they are developing learner characteristics that they cannot if they are passive participants in a classroom. While planning such engaging lessons, tasks and units can be hard work, it is ultimately worth the effort every time.

Introducing key concept 1: form

	ATL Skills	Mathematics Skills
Activity 1 Forms of a quadratic function	✓ Organize and depict information logically.	**Algebra** ✓ Rewrite quadratic functions in various forms.
Activity 2 Solving quadratic equations	✓ Organize and depict information logically.	**Algebra** ✓ Solve quadratic equations in various ways. ✓ Solve a real-world problem.
Activity 3 Framing a photograph	✓ Consider content.	**Algebra** ✓ Rewrite quadratic functions in various forms. ✓ Solve quadratic equations in various ways. ✓ Solve a real-world problem.

Introducing form

Every new body of discovery is mathematical in form, because there is no other guidance we can have.

Charles Darwin

The key concept of form in MYP mathematics refers to the understanding that the underlying structure and shape of an entity is distinguished by its properties. The study of form helps students understand the different ways in which mathematical ideas and relationships can be expressed. There are different forms for representing the equation of a line, just as there are different methods of organizing and representing data. The study of form not only looks at these but also allows students to explore the advantages and disadvantages of choosing to express something in a particular way. In this chapter, students will develop the concept of form by looking at the different ways of representing a quadratic relationship, something that they will continue to do in future chapters.

Activity 1 Forms of a quadratic function

In this activity, students are asked to write quadratic functions in three different forms and to translate from one form to another, so that they are familiar with them. Note that, in each case, $a \neq 0$.

Standard form: $y = ax^2 + bx + c$

Factorized form: $y = a(bx + c)(dx + e)$

Vertex form: $y = a(x - h)^2 + k$

TEACHING IDEA 1: Effects of parameters

INQUIRY The web links box in the student book (http://www.mathopenref.com) takes students to a page where they can see the effects of varying the parameters on the standard and vertex forms of a quadratic function. Rather than telling students what the effects are and then verifying them by referring to the site, encourage them to generalize the effects after experimenting with the graphs and sliders on the webpage.

TEACHING IDEA 2: Completing the square

INQUIRY Some students find it very difficult to convert a quadratic function in standard form into its equivalent in vertex form by completing the square. Instead of teaching the process directly, encourage your students to develop the technique for themselves, by working through the exercise below.

This short exercise is intended to ensure that students are comfortable with squaring binomials and that they have had some prior work with perfect square trinomials.

WEB LINKS

For a visual representation of the procedure for completing the square, visit http://illuminations.nctm.org and search for "proof without words: completing the square".

a) Expand each expression. Show **all** of your working.

(i) $(x + 4)^2 - 10$ (ii) $(x - 1)^2 + 3$
(iii) $(x + 8)^2 + 1$ (iv) $(x - 3)^2 - 5$
(v) $(x - 2)^2 + 8$ (vi) $(x + 5)^2 - 30$
(vii) $(x + 1)^2 - 7$ (viii) $(x - 6)^2 - 22$

b) How do the coefficients of each resulting trinomial relate to the values in the original question?

c) Choose a value of c to make each of these trinomials into a perfect square: $(x - h)^2$. Use fractions instead of decimals.

(i) $x^2 + 4x + c$ (ii) $x^2 - 12x + c$
(iii) $x^2 + 24x + c$ (iv) $x^2 - 5x + c$

d) Now rewrite each of these quadratics in the form $(x - h)^2 + k$. This is called "completing the square".

(i) $x^2 + 4x - 7$ (ii) $x^2 - 14x + 1$
(iii) $x^2 + 6x - 5$ (iv) $x^2 - 20x + 10$

e) Describe how to complete the square of a quadratic function in standard form: $y = x^2 + bx + c$.

Activity 2 Solving quadratic equations

In this activity, students will use three different methods to solve quadratic equations (factorizing, quadratic formula, completing the square). Students should discover that, depending on the form of the equation, some methods might be preferable to others. Once they understand the three methods, they should be able to write the flow chart to answer question **b**).

TEACHING IDEA 3: The quadratic formula

INQUIRY If students are comfortable with the method of completing the square, then they could be asked to complete the square of a quadratic equation in standard form and discover the quadratic formula for themselves.

 Activity 3 Framing a photograph

Students are given the following problem.

Grandfather is planning to dismantle an old treehouse, but before he does, he decides to keep a framed photo of it. He will enlarge the picture to 200 mm × 150 mm and use some wood from the treehouse to frame his photo.

After dismantling the treehouse, he decides to use a piece of wood that measures 255 mm × 125 mm. He wants to use the whole piece of wood for the frame. How wide does the frame need to be? The frame should be the same width, all the way around the picture, and no pieces should overlap. He may have to cut the piece of wood into more than four pieces before he assembles it to make the frame.

TEACHING IDEA 4: Introducing the activity

An engaging way to introduce the activity is to give students a photo with the dimensions in the problem, as well as a piece of paper that is the same size as the reclaimed piece of wood for the frame. Let them attempt to cut the "wood" and assemble it to try to make the frame. When they see how difficult it is to do this, they will be encouraged to find a solution. Once they have found the solution mathematically, provide them with another piece of "wood" so they can work out how to cut it and, at the same time, verify their answer!

One of the goals of this activity is to help students recognize the value in having different methods to solve a problem. However, each one has its distinct advantages, making it easier to apply in some situations.

⊂⊃ INTERDISCIPLINARY LINKS
Sciences
Analysing the motion of objects in physics often involves solving quadratic equations. Some of those equations can be factorized while others require the use of the quadratic formula.

TIP

The quadratic equation that results from the activity is $4x^2 + 700x - 31\,875 = 0$. This can be factorized to give $(2x - 75)(2x + 425) = 0$. Students can also solve it by using the quadratic formula.

Summary

The key concept of form underlies many topics in mathematics, so it is important that students are aware of it. An understanding of form leads to appropriate choices for representing quantities, relationships or information and the advantages and disadvantages of any particular representation. It is fundamental to many mathematical concepts such as structure, logical argument, proof and comparison.

Introducing key concept 2: logic

	ATL Skills	Mathematics Skills
Activity 1 Solving Calcudoku puzzles	✓ Demonstrate persistence and perseverence.	**Number** ✓ Use logic to solve number puzzles.
Activity 2 Master tips	✓ Listen actively to other perspectives and ideas.	**Number** ✓ Use logic to solve number puzzles.
Activity 3 Angle relationships	✓ Consider personal learning strategies.	**Geometry and trigonometry** ✓ Find the size of an unknown angle.

Introducing logic

In MYP mathematics, the key concept of logic refers to the process used in making decisions about numbers, shapes and variables. It is important that students understand that logic is a process that provides certainty. Students use logical systems of reasoning to draw conclusions and explain their validity. Logic is present in all four branches of mathematics, from determining the size of angles in a diagram to simplifying algebraic expressions. As they work through the activities in this book, students will use logic, for example, to learn new mathematical concepts based on patterns they recognize among numbers, shapes and expressions. It is an inescapable part of learning and applying mathematics, since it helps students to draw conclusions consistent with what they observe and calculate.

Pure mathematics is, in its way, the poetry of logical ideas.

Albert Einstein

As they start these activities students may work individually or in pairs. Working with a partner encourages dialogue and skills in communicating mathematics verbally. As students work through the puzzles, which get progressively more difficult, they will record the problem-solving strategies that helped them solve each puzzle.

Remind students that the solution to any Calcudoku puzzle is unique.

These are the solutions.

Grid 1 (3 × 3)

3+	1 −	3
1	**2**	**3**
2	**3**	2÷ **1**
3÷ **3**	**1**	**2**

Grid 2 (4 × 4)

6 ×			4
2	**1**	**3**	**4**
12 × **1**	1 − **3**	**4**	5 + **2**
3	2÷ **4**	**2**	**1**
4	**2**	2 − **1**	**3**

Grid 3 (5 × 5)

8 ×	1 −	4 −	2 −	
4	**2**	**1**	**3**	**5**
2	**3**	**5**	10 + **1**	**4**
1	1 − **4**	**3**	**5**	2 **2**
2 − **3**	**5**	2 ÷ **4**	**2**	2 − **1**
10 × **5**	**1**	**2**	4 **4**	**3**

🔗 **WEB LINKS**
Calcudoku puzzles are available at http://www.kenken.com. The site has excellent resources for teachers, problem-solving tips and online puzzles that you can create and customize to cater for any difficulty level. Once the students have completed the puzzles in this book (which were created from this website), you might encourage them to try a more complicated one online. Tip—you can turn off the Auto Notes so that the students must work through the process of elimination of possible numbers for a particular square, but students can reveal squares or see the full solution if they wish. If you want the students to work through the whole process you can print out the KenKen puzzles you create, instead.

Ask the students, in pairs, to summarize their strategies and tips. Then let each pair join up with another pair and share their ideas. As a group of four, they now write out their strategies, ready to share with the class.

 **Activity 2** | **Master tips**

Display an example that the class can work through together. You could use the website http://www.kenken.com and either print a new puzzle or project one on screen. Ask each group to explain one of their strategies, using the current puzzle to demonstrate it.

Students will suggest many ideas and strategies on how best to solve this type of logic puzzle but the following ideas should be unanimous.

- Start by filling in boxes that have only one possible value, such as single boxes.
- Write down all possible combinations of numbers in each cage.
- Try to fill in combinations of numbers that have the fewest possible combinations.
- Look at the factors of numbers and possible combinations, given the different operations and divisibility patterns.
- Use the rule that no number can appear twice in any row or column to help with the process of elimination—if a square contains a possible number that is found nowhere else in that row or column then it clearly must be that number.

Calcudoku puzzles are just one type of number puzzle that students can enjoy. Many students may already be familiar with the popular Sudoku puzzles that use similar strategies but in a different game. There are now even Paintdoku puzzles, in which the player uses logical reasoning to decide what boxes to shade on a massive grid. At the end of the game, the player has "drawn", or shaded, a picture.

Below are the solutions to the more challenging puzzles.

WEB LINKS

Details about each of the different types of puzzle mentioned here can be found at Archimedes Laboratory, http://www.archimedes-lab.org. Simply search for the type of puzzle you want.

INTERDISCIPLINARY LINKS

Arts

If students enjoy the Paintdoku puzzles, they could be asked to create their own puzzle as an interdisciplinary activity with their arts class.

Puzzle 1

1 − 5	4	3 + 2	1	1 − 3	6 × 6
3 ÷ 3	11 + 6	5	24 × 4	2	1
1	8 + 2	3	6	80 × 4	3 − 5
2 − 6	3	4	5	1	2
4	5 − 1	6	10 + 2	5	3
10 × 2	5	3 ÷ 1	3	2 − 6	4

Puzzle 2

24 × 6	4	13 + 5	3	2 − 2	8 + 1
5 − 1	6	2 ÷ 3	5	4	2
120 × 4	1 − 2	6	4 + 1	2 ÷ 3	5
5	3	1	2	6	13 + 4
3	12 + 1	2	20 × 4	5	6
2	5	4	5 − 6	1	3

Grid 1

3 ÷ **1**	336 × **4**	**7**	1 − **2**	2 − **5**	**3**	1 − **6**
3	**2**	**6**	**1**	17 + **4**	**7**	**5**
6 − **7**	**1**	24 × **2**	1 − **5**	**6**	1 − **4**	**3**
17 + **2**	**3**	**4**	**6**	35 × **1**	**5**	**7**
6	**5**	21 × **3**	**7**	4 × **2**	**1**	3 − **4**
4	11 + **6**	**5**	14 + **3**	**7**	**2**	**1**
5 **5**	6 − **7**	**1**	**4**	36 × **3**	**6**	**2**

Grid 2

210 × **5**	72 × **4**	**3**	**1**	11 + **7**	4 − **2**	**6**
7	105 × **3**	**6**	5 **5**	**4**	2 ÷ **1**	**2**
6	**5**	**7**	1 − **3**	**2**	3 − **4**	**1**
13 + **4**	2 ÷ **1**	**2**	18 + **6**	**5**	**7**	15 + **3**
2	**7**	15 + **1**	**4**	3 − **6**	**3**	**5**
6 × **3**	**6**	**4**	14 + **2**	**1**	**5**	**7**
1	**2**	15 + **5**	**7**	**3**	**6**	4 **4**

TEACHING IDEA 1:

Do it yourself

Let students create their own Calcudoku or Sudoku puzzles and then swap them. Begin with 3 by 3 puzzles and then challenge them to create their own larger puzzles.

EXTENSION

a) Ask students whether they think it is possible to create a 2 by 2 Calcudoku puzzle in which none of the numbers is given explicitly. Ask them to use logical argument to prove their answers.

b) "No operation" KenKen-style puzzles have now been developed. These are similar to the above examples but the operation in each cage is hidden. How does this alter the difficulty level? Go to http://www.mathdoku.com and select a no operation game to try one.

Activity 3 | **Angle relationships**

In this activity, students are using their knowledge of the relationships between angles in triangles as well as between those formed by lines and transversals to determine the size of unknown angles. In order to complete the two puzzles, students will need to be familiar with the following facts:

- the sum of the interior angles of a triangle is 180 degrees (180°)
- the angles of an equilateral triangle are all equal
- the angles opposite equal sides of an isosceles triangle are equal
- alternate interior angles are equal
- vertically opposite angles are equal
- corresponding angles are equal.

TIP

Encourage students to create a set of strategies for solving these "angle puzzles" and compare them to those from the Calcudoku activity. Students should realize (or be shown) that understanding and applying "the rules" is the key to success in solving most problems.

TEACHING IDEA 2:
Transversals

 Students can discover the angle facts through an inquiry method. For example, they could be asked to draw two parallel lines and a transversal and then use a protractor to find which pairs of angles are equal. They could also use dynamic geometry software to do this. They could then either be given the appropriate vocabulary or asked to research what these pairs of angles are called. Finally, students could be asked to justify or prove their results formally.

Summary

Logic is a method of reasoning that is used throughout mathematics, including in puzzles that may otherwise seem to lack mathematical content. However, the same logic that is used to solve puzzles can also be applied to most of the topics encountered in a mathematics classroom. The skills students develop in looking for patterns or relationships and establishing general rules are invaluable in all mathematics courses. As students work through this book, encourage them to realize how often they are using logic, either to solve a problem or generate a rule. Certain problems may, in fact, seem more like puzzles to students once they have the prerequisite knowledge. Combined with good basic skills, logic provides a means of acquiring, generalizing and applying mathematical concepts.

TIP

Some students find visual clues helpful, to remind them of the different angle relationships.

- Vertically opposite angles form an X.
- Corresponding angles form an F.
- Alternate interior angles form a Z.
- Interior angles on the same side of a transversal form a C.

As they solve missing-angle puzzles, they can look for these letter-shapes within the problem.

Introducing key concept 3: relationships

	ATL Skills	Mathematics Skills
Activity 1 Mathematical relationships in medicine – avoiding too much radiation	✓ Organize and depict information logically.	**Geometry and trigonometry** ✓ Solve proportions that arise from similar triangles. ✓ Establish relationships between lengths of sides of triangles.
Activity 2 Mathematical relationships in nature – where do babies come from?	✓ Draw reasonable conclusions and generalizations.	**Statistics** ✓ Use a GDC to graph data and perform a regression analysis. ✓ Interpret the relationship between two variables.

Introducing relationships

The concept of relationships in mathematics is fundamental and encompasses many different ideas and topics. This key concept refers to the connections between quantities, properties or, more often, variables, from the relationship between decimals, ratios and percentages to how two variables relate to one another and beyond. These relationships can then be expressed as a model, a rule or a mathematical statement. Establishing such relationships is fundamental if students are to be able to look at the world around them and describe the patterns that they see in a meaningful way. These relationships also enable students to determine new relationships based on those that they already know. Every branch of mathematics: number, algebra, statistics and probability, and geometry and trigonometry, involves establishing relationships and then using them to make predictions or determine unknowns. In this chapter students will see examples of what can result when they can identify and/or apply such relationships.

> *Mathematicians do not study objects, but relations between objects.*
>
> Henri Poincaré

TIP

Many students think that formulas and relationships are separate entities. Asking questions such as "What is the relationship between the area of a circle and its radius?" as opposed to "What is the area of a circle?" is a subtle way to remind students that formulas express relationships between quantities and/or measurements.

QUICK THINK

The quick think activity leads students to think back to what they have learned so far and see what relationships were established in each branch of mathematics. Typical responses for relationships that can be expressed as an equation or formula are: slope, midpoint, distance between points, area and volume formulas, circumference of a circle, mean and linear functions. Typical answers for relationships that are not given in the form of an equation or formula include the relationship between equivalent fractions, decimals and percentages.

Activity 1 Mathematical relationships in medicine—avoiding too much radiation

In this activity, students will first apply the concept of similarity to a medical application. To be successful, students should know:

- how to determine if two triangles are similar
- how to justify that two triangles are similar
- that pairs of corresponding sides of similar triangles are in the same proportion.

TEACHING IDEA 1: Similar triangles

INQUIRY For students who do not know the properties of similar triangles, or who need to review them, the web links box in the student book (www.mathopenref.com) provides an interactive way of developing the ideas of proportionality of sides in similar triangles. After interacting with the site, students can be asked to think of their own definitions and descriptions of the properties of similarity.

Students will first need to label their diagram. One possibility is shown below. Since corresponding sides are proportional, students should be able to express that:

$$\frac{AC}{AB} = \frac{AD}{AE} = \frac{CD}{BE}$$

Students should then be able to use this relationship to find that:

$$\frac{80}{75} = \frac{x}{10}$$

Solving this, they should find the value of x to be 10.67 cm. In order to avoid too much radiation, the sources should be at least double this distance (21.34 cm) apart.

Students are asked to reflect on the fact that the radiation will actually emanate from the sources as a cone, rather than a triangle. They are then asked to state any assumptions that they have made with regard to the problem and to analyse the effect that it may have on the solution. One possible response could be that it is assumed that the radiation does not penetrate past the spinal cord. If it does, then anything beneath it may get a double dose of radiation.

TIP

Students' natural inclination is to compare the two large triangles and state that they are similar to each other. It is a good idea to ask them how they established this. Is it enough that they simply **look** as if they are the same shape? The activity requires students to find the similar triangles within **one** of the larger ones. In fact, there are two sets of similar triangles, both of which may be used to solve the problem. Students could be asked to solve the problem with both pairs.

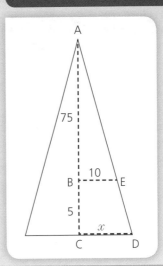

REFLECTION

Encourage students to discuss the points raised in the student book. These are real problems that have to be faced in the world of medicine.

Activity 2   Mathematical relationships in nature—where do babies come from?

In this activity, students will explore the relationship between two variables and describe it mathematically. They will need to know how to draw a graph and find the equation of a line, either by hand or using a GDC.

⬯ INTERNATIONAL MATHEMATICS

There are many cultural variations on the answer to the age-old question, "Where do babies come from?" In France, one such answer is *"les filles sont nées dans les roses et les garçons dans les choux"* ("girls are born in the roses and boys in the cabbages"). Data could be found for the number of births and rose or cabbage production to arrive at the same conclusion. Students could even be asked to find data related to their own cultural or traditional explanation.

Students may have difficulty in trying to suggest appropriate explanations for the simultaneous increase and decrease of the two populations. It is most important that they realize that there may be an entirely different factor that has an effect on the variables being studied. Ultimately, it is important for students to learn that, even though a relationship may appear to exist between two variables, statistics do not prove that a change in one variable has caused a change in the other. Even when several sources of information point to the existence of a relationship, care must be taken when drawing conclusions. It may be appropriate to introduce the idea of a confounding variable, towards the end of the discussion.

REFLECTION

Students are asked to discuss the assumptions and conclusions they have made. Encourage them to consider whether there may be other factors that they have not considered.

> **TIP**
>
> The idea of storks bringing babies is a cultural one. Many students may never have heard of this legend. Therefore, they may need to be introduced to it beforehand, for the choice of data to have any meaning.

> **TIP**
>
> This activity could be used at the beginning of a unit on either functions or statistics. If students have access to a GDC, then they can use it to complete the entire task, including answering all questions. If not, they can easily complete parts a) and b) but may need some guidance on how to justify their choice of a model.

Summary

Relationships are an integral part of MYP mathematics and can be found in almost every unit of study. As students develop the ability to establish relationships between quantities, properties and concepts they become more adept at recognizing patterns and constructing generalizations. As students learn to communicate relationships in a mathematical way, they are better able to use correct mathematical language and notation, skills that also help them in communicating any mathematical thinking. The skills students acquire in studying, establishing and applying mathematical relationships are at the heart of much of what they need to be able to do in MYP mathematics.

Change
A variation in size, amount or behaviour

	ATL Skills	Mathematics Skills
TOPIC 1 Finite patterns of change		
Activity 1 Playing with numbers and exploring arithmetic series	✓ Understand and use mathematical notation.	**Algebra** ✓ Identify arithmetic and geometric sequences and series.
Activity 2 How do I know how far I can go?	✓ Help others to succeed.	**Number** ✓ Use standard form or scientific notation to represent numbers. ✓ Simplify results and round answers to an appropriate level of accuracy. **Algebra** ✓ Identify arithmetic and geometric sequences and series. ✓ Use diagrams to represent information and solve problems.
TOPIC 2 Infinite patterns of change		
Activity 3 Calculating the probability of winning an endless game	✓ Evaluate and manage risk.	**Number** ✓ Simplify results and round answers to an appropriate level of accuracy. **Algebra** ✓ Identify arithmetic and geometric sequences and series. **Probability** ✓ Calculate the probability of more than one outcome for independent events.
TOPIC 3 Financial mathematics		
Activities 4–6 Investing in the stock market, bonds and a savings account	✓ Identify trends and forecast possibilities.	**Number** ✓ Calculate percentage increase and decrease. ✓ Simplify results and round answers to an appropriate level of accuracy. **Algebra** ✓ Calculate investment returns, using simple interest.
Activity 7 Stock market trends and volatility	✓ Identify trends and forecast possibilities.	**Number** ✓ Calculate percentage increase and decrease. ✓ Simplify results and round answers to an appropriate level of accuracy.
OTHER RELATED CONCEPTS	Pattern Generalization Quantity Representation	

Introducing change

Change represents a variation in size, amount or behaviour. In this chapter, students will discover different patterns of change, explore them in a variety of mathematical and real-life contexts and reflect on how the knowledge of these patterns can help them to make informed decisions when making investments, borrowing money or simply playing games.

At the beginning of the chapter, students are introduced to two legends—the invention of the game of chess and the legend of the Hanoi Towers. Both highlight how quickly the terms of a geometric sequence (also called geometric progression) can increase and, when consecutive terms are added, the totals become incredibly large numbers.

> *To improve is to change; to be perfect is to change often.*
>
> Winston Churchill

TOPIC 1　Finite patterns of change

At the beginning of this topic students can read about Carl Friedrich Gauss and how, as a small boy, he discovered an amazing method to add consecutive numbers quickly. Gauss is considered one of the foremost mathematicians of all time, due to his contributions to a wide variety of fields in both mathematics and physics (including number theory, analysis, differential geometry, geodesy, magnetism, astronomy and optics). It is worth asking the students to learn more about his work.

The activities in the student book deal with arithmetic series, finding the sum of consecutive terms of arithmetic sequences (or progressions). In case students are unfamiliar with this concept, the first teaching idea provides a nice introduction to the topic.

TEACHING IDEA 1: Make a presentation
Divide students into groups and get them to prepare a class presentation using different media or simply make posters about the importance of Gauss' contributions to mathematics and science in general.

TEACHING IDEA 2: Summer Olympics and FIFA World Cup

This table gives the years of the Summer Olympic games since 1980.

1980	Moscow, USSR
1984	Los Angeles, United States
1988	Seoul, South Korea
1992	Barcelona, Spain
1996	Atlanta, United States
2000	Sydney, Australia
2004	Athens, Greece
2008	Beijing, China
2012	London, UK

Ask students to deduce a formula for the nth term of the sequence of years when the summer Olympic games took place, starting in 1980. They can research and verify that the formula works.

Students could then do some research and modify their formula to give the years when the FIFA World Cup took place.

Instead of giving the table with the years of the Summer Olympic games, you could let the students research the topic and learn more about the Winter and Summer games, when they started and whether or not the pattern was followed all the time.

Activity 1 **Playing with numbers and exploring arithmetic series**

This investigation provides an excellent opportunity for students to discover the relation between the middle term and the sum of n consecutive terms of an arithmetic sequence when n is odd. Students need to complete a table and are led to deduce that:

1. $2 \times middle\ term = a_1 + a_n$
2. $a_1 + a_2 + \cdots + a_n = middle\ term \times n$

EXTENSION

Students can investigate further the sum of consecutive numbers by considering:

- series with an even number of terms
- series where $a_1 > 1$.

As the students investigate other arithmetic series and explore examples where n is even, they need to refine their conclusions:

3. $sum\ of\ middle\ terms = a_1 + a_n$

4. $a_1 + a_2 + \cdots + a_n = \dfrac{a_1 + a_n}{2} \times n$

In the last part of the activity, students are guided to use two different ways to prove the result they have discovered. They are also given the opportunity to apply their findings and calculate the sum of several arithmetic series, including series with irrational terms. This can be a good opportunity to revise simplification of surds (radicals). For example, students can work in pairs or small groups and complete this activity.

a) Shown below are radicals and their simplified forms. Determine the pattern that would allow you to simplify a radical completely.

$\sqrt{24} = 2\sqrt{6}$ $\sqrt{18} = 3\sqrt{2}$ $\sqrt{32} = 4\sqrt{2}$ $\sqrt{300} = 10\sqrt{3}$

$\sqrt{20} = 2\sqrt{5}$ $\sqrt{48} = 4\sqrt{3}$ $\sqrt{27} = 3\sqrt{3}$ $\sqrt{125} = 5\sqrt{5}$

b) Use your calculator in each case to check if your rule works.

c) Use your rule to simplify these radicals completely.

$\sqrt{28} = $ _____ $\sqrt{45} = $ _____ $\sqrt{72} = $ _____ $\sqrt{64} = $ _____

d) Modify your rule to simplify these radicals.

$\sqrt[3]{54} = $ _____ $\sqrt[4]{32} = $ _____ $\sqrt[3]{56} = $ _____ $\sqrt[3]{640} = $ _____

e) Describe the general rule to simplify radicals clearly. Use correct mathematical notation. Provide further examples that you consider relevant to clarify the application of this rule.

TEACHING IDEA 3: Try it and see

After they have read about the Hanoi Towers, let the students use a model or an interactive simulator to try for themselves to move a collection of discs from one pin to another, following the rules of the game. They can then discover a systematic method, using as few movements as possible. For n discs the minimum number of movements is $2^n - 1$ which, for $n = 64$, gives exactly the value that the chess inventor would get as a reward for his accomplishment.

 DP LINKS

Arithmetic sequences are explored further in mathematics in the IB Diploma Programme because a variety of phenomena change in this way.

Exploring the limits of a process

The second part of Topic 1 focuses on exploring geometric series. To make the nature of exponential growth obvious, students are challenged to build paper towers (stacks of paper) and measure their dimensions. Then they predict the amount of paper needed to build a tower with the same height as a real skyscraper or tower. Finally, they are asked to reflect on the limitations of the process.

During the activity, students will deal with both large and small numbers. This activity offers a good opportunity to introduce or review standard form (also called scientific form or scientific notation) and talk about the advantages of its use.

 WEB LINKS

A fascinating application was created by a set of twin brothers while in the 9th grade. Let students visit http://htwins.net/scale2 to see their project: it demonstrates examples of objects over an incredible range of powers of 10.

TEACHING IDEA 4: Standard form

INQUIRY Ask students to research standard form, its advantages and subjects where it is used frequently.

Research and make a summary of operations in standard form, that is, how you add, subtract, multiply, divide and find powers of values expressed in standard form.

Activity 2 — How do I know how far I can go?

This group task is broken into two parts. In task 1, students need to use a large number of paper sheets in order to estimate their thickness. Alternatively, the value of the thickness can be given to students. Before they start the activity it is important that students are familiar with standard form and know how to store values in their calculators. Instructions on how to do this when using a TI-Nspire are available on page 27 of the student book.

Task 2 requires students to investigate what happens to the number of layers and the thickness as a sheet of paper is folded multiple times. It is here that they will discover a geometric sequence and then reflect on the feasibility of creating a "stack" (or a tower) by folding paper.

Activity 2 also involves dealing with accuracy and estimation. As the values increase or decrease rapidly, students need to consider the level of accuracy carefully.

The reflection task at the end of Topic 1 requires students to explore geometric sequences and series further, and consider the results obtained. The students apply their learning and compare an arithmetic model with a geometric model and decide which investment option is the best.

Assessment

This reflection task can be adapted to be assessed against criteria C and D. It can also be modified to make it suitable to be assessed against criterion B. In this case, the task needs to include the opportunity for students to discover the general rules to determine the total interest gained after n years for an interest rate r for each of the options given.

TEACHING IDEA 5: Research activity

INQUIRY As this topic deals with some huge numbers and some infinitely small numbers, it can be interesting for students to research terminology used for huge numbers and read about who invented them and how—what is a googol and a googolplex? Do these quantities really exist? Can they write down all the digits of a googol? If so, how long does it take?

DP LINKS

Geometric sequences and series are studied in detail in mathematics in the IB Diploma Programme. This includes their applications to financial mathematics that involve compound interest applications.

MATHS THROUGH HISTORY

Geometric progressions have been found on Babylonian tablets dating back to 2100 BC. Arithmetic progressions were first found in the Ahmes Papyrus, which is dated at 1550 BC.

Infinite patterns of change

Activity 3 requires students to investigate an old dice game that has many names. In the US it is called Rolling Dice or Craps. In order to be successful, students will need to be familiar with calculating the probability of two or more independent events.

TEACHING IDEA 6: Deducing the sum of an infinite geometric series

INQUIRY Before assigning this activity, students might benefit from tackling a simpler problem on the idea of an endless game. The following is a suggestion as an introductory "endless game".

Abe flips two fair coins together, and continues to do so until at least one of the coins turns up heads. Find the probability that both coins turn up heads on the last flip.

Students will quickly realize that this game could theoretically go on forever, since the coins could always turn up tails. He will always turn up TT, HT, TH or HH, each with a probability of $\frac{1}{4}$. To succeed, Abe must either get HH on the first throw, or get TT successively a number of times before finally getting HH. (He can't get an HT or TH as then the game is over, and he won't have the possibility of getting HH.)

To win on the first throw, he needs HH, which has a probability of $\frac{1}{4}$ of occurring.

To win on the second throw, he needs to first get TT, followed by HH, that is, a probability of $\frac{1}{4} \times \frac{1}{4}$, or $\frac{1}{16}$.

To win on the third throw, he needs to first get TT, then again TT, followed by HH, or $\frac{1}{4} \times \frac{1}{4} \times \frac{1}{4}$ or $\frac{1}{64}$.

Therefore, to win on the first, second or third throw, the students add the probabilities to obtain $\frac{1}{4} + \frac{1}{16} + \frac{1}{64} = \frac{21}{64}$ (or approximately 0.328).

Students should be able to conjecture, therefore, that to win on the first, second or nth throw, they add $\frac{1}{4} + \left(\frac{1}{4}\right)^2 + \left(\frac{1}{4}\right)^3 + \cdots + \left(\frac{1}{4}\right)^n$. They should notice that this is a geometric series, and can use the formula to get an expression in n, that is,

$$\frac{\frac{1}{4}\left(1 - \left(\frac{1}{4}\right)^n\right)}{1 - \frac{1}{4}}.$$

From this formula, invite students to think about what happens as the number of throws gets larger and larger, that is, as n approaches infinity. Then $\left(\frac{1}{4}\right)^n$ approaches 0, and the probability that the game is endlessly long is $\frac{1}{3}$.

The formula for an infinite geometric series can thus be deduced, as well as the condition that $|r| < 1$.

Assessment

If you choose to assess students on this task, you can use criterion D. The task-specific descriptor in the top band (7–8) should read that the student is able to:

- **identify** all relevant rules for playing the game
- **select** appropriate mathematical strategies to analyse all relevant elements of the game
- **apply** the mathematical strategies to correctly calculate all probabilities
- **justify** the degree of accuracy of their solution
- **justify** their result in a real-world context.

Stage 1 of the unit planner

Key concept	Related concepts	Global context
Logic	Change Justification	Identities and relationships
Statement of inquiry		

Logic can be used to justify why we choose to engage in or avoid risky behaviours.

In this activity, students model some of the axioms of probability theory, using a real-life dice game. It is very important that students understand the rules of the game, so spend some time at the beginning of the lesson going over them. A few rolls of the dice before they actually begin the activity should help to clarify any questions students may have.

In this activity, students first create a grid like the one on page 30 of the student book. In steps 2–5 they will interpret their results from the grid.

Answers

For steps 6–10, students will need to use the fact that the probability of two or more independent outcomes is the product of the probabilities of the outcomes.

The answer to the question in step 8 is:

$$P(5, \text{then not 5 and not 7, then not 5 and not 7, then not 5 and not 7, then 5}) = \frac{4}{36} \times \frac{26}{36} \times \frac{26}{36} \times \frac{26}{36} \times \frac{4}{36} = \left(\frac{1}{9}\right)^2 \times \left(\frac{13}{18}\right)^3$$

If students have worked carefully and can fill out the table successfully, they should arrive at the formula for step 10,

$$P(\text{winning on the } n\text{th throw}) = \left(\frac{1}{9}\right)^2 \times \left(\frac{13}{18}\right)^{n-2}.$$

In step 11, students will have to add their results from the table. This means that they find the probability of scoring a 5 on the first and second throws, or on the first, second and third throws, and so on. They can then use the formula for finding the sum of the first n terms of a geometric series, and either deduce the probability for winning after an endless number of throws, or use the formula for the sum of an infinite geometric series, since $r = \frac{1}{4}$.

For step 12, students will now do the same as above for the different points that can be scored on the first throw. In step 13, they will then compare the probabilities obtained for the different points, and determine which one is greatest.

For step 14, students will need to consider all the different ways of winning the game, and add these probabilities. Alternatively, they can use the probability axiom $P(A) = 1 - P(A')$. This is indeed far easier, and some clever students may spot this right at the beginning!

TOPIC 3 · Financial mathematics

Clearly, a lot of students will see the relevance of this topic immediately and will be engaged when looking at anything to do with making money.

The focus of the mathematics is linear versus exponential functions, so it is important that all students have some understanding of these before they attempt the tasks in this topic.

TEACHING IDEA 7: What's your worth?

INQUIRY As you will teach linear functions before exponential functions, here is an interesting introductory activity. Ask the students to choose between being paid $2000 per day for one month, or start with 1 penny and have the amount doubled each day, so each payment is exactly double the payment of the day before. (You can change the currency units to whatever you like.) Some students jump at the $2000 a day, while others are suspicious and assume that there is a trick here. Let them create a table in a spreadsheet to record the cumulative amount of money received each day, for the two options, and then graph the results.

You can use this investigation to discuss the nature of exponential functions and the comparisons to linear functions. As this is a very basic example of an exponential function, students should be able to recognize that the graph of the exponential growth function becomes very steep quickly while the linear function increases at a constant rate and hence why the "penny doubled each day" option is better. You could then discuss the legend about the grand vizier in Persia, which is mentioned at the beginning of this chapter in the student book.

TIP

By setting up the formulas in a spreadsheet or using TI-Nspire which includes spreadsheet features, the students can simply fill the formulas down the columns then create the graph.

CHAPTER LINKS

If you choose to have the financial mathematics task as your summative task at the end of an exponential unit, then the exponential growth and decay activities in Chapter 11 on Patterns will work well as an investigation during the unit.

Activities 4–6 · Investing in the stock market, bonds and a savings account

Assessment

If you choose to assess students on this task, you can use criterion D.
The task-specific descriptor in the top band (7–8) should read that the student is able to:

- **identify** <u>all</u> relevant elements of <u>all</u> three investment options (stocks, bonds and bank account scenarios)
- **select** appropriate mathematical strategies to analyse <u>all</u> three investment options
- **apply** the selected mathematical strategies and correctly calculate <u>all</u> possible investment outcomes
- **justify** the degree of accuracy used in the calculations
- make a <u>comprehensive</u> comparison of the three investment options with a <u>detailed</u> **justification** for the preferred investment choice.

Key concept	Related concepts	Global context
Relationships	Change Patterns Models	Identities and relationships
Statement of inquiry		
Representing patterns of change as relationships can help us make better personal choices.		

The aim of this activity is for students to calculate the percentage gain or loss that they could make on the stock market by looking at the price of the stocks when they bought them, compared to when they sold them.

Students may need help to find the current and historical prices on their index. You may even decide to tell them which index to use, as some students might find the idea of finding the correct index daunting.

EXTENSION

Linear regression—stock market trends

A useful introduction to the stock market is to look at one of the indices over a large period of time and let students interpret how the market has behaved. While it is normal to look at trends over time and, typically, use linear regression to show the general trend, the goal is for students to see that the market does not behave in a linear predictable pattern, as there are so many factors that influence stock price and market behaviour. You could let the students complete the linear regression themselves and check it against their results from this activity. Alternatively, different groups of students could use different indices to compare different countries or types of stocks—all good points for discussion.

Stock price setting—supply and demand

As the stock price is determined by the supply and demand for the stock (a linear system), you could suggest an investigation into how this occurs. Start with a basic linear system.

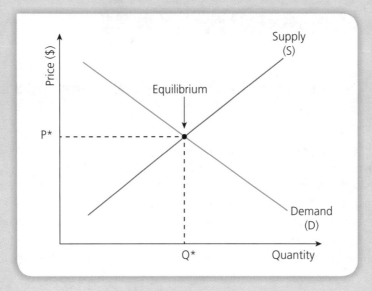

There is no need to label any scale on the axes as the specific numbers are not important—just make clear that quantity is on the horizontal axis and price is on the vertical axis. Once you add in the new shifted demand and supply lines, it is the direction of the shift and new equilibrium points in relative position to the old ones that are the focus. Ask students what will happen to the demand line if demand increases. Draw in the new line and show how the equilibrium point has shifted. The same can be done with a decrease in demand. Repeat the process for shifts in the supply line.

When looking at stock prices, an increase in demand (people wanting to buy the stock) tends to coincide with a decrease in supply (people willing to sell the stock) so both lines shift to create a new higher stock price. The opposite is also true.

Try giving these scenarios to the students to see if they have grasped the concept.

- A company's yearly income result is less than their forecasted income—supply increases and demand decreases.
- A company that makes a drug gets approved by the FDA (or equivalent organization in your country)—supply decreases and demand increases.

You can even discuss other events that would alter stock price and see how it would affect the demand/supply graph.

Investing in bonds

The goal of this activity is for students to calculate how much money would be made if the money was invested in bonds. To do this, they create a linear function to represent this change over time. They will find that, generally, while they might not make as much money in government bonds as they can on the stock market, they are a safer and more reliable investment.

Investing in a savings account

Students will calculate how much money would be made if the money was invested in a savings account. To do this, they will create an exponential function to represent this change over time. They will then compare the linear graph representing investing in bonds with the exponential graph of the savings account.

Students will need to research interest rates and compounding periods of different savings accounts offered by their local banks. Alternatively, you could gather that information for them. It would be useful if students have had practice answering basic questions involving savings accounts prior to this summative task, to ensure they understand all the components of this formula. You could let them carry out an investigation on how different interest rates and different compounding periods affect the exponential graph and interest earned.

WEB LINKS

Interactive models can be found at: http://reffonomics.com. Select **BasicConcepts>Supply and Demand**. In particular look at what happens when the shifts in supply and demand occur at the same time.

DP LINKS

In Diploma Programme economics, students study supply and demand in great detail. Not labelling any scales on either axes makes this activity a nice introduction as it is only looking at shifts in demand and supply and no further analysis is needed for the students to grasp the concept of how price is determined.

TIP

When the students are deciding where to invest their money, there is no right or wrong answer as it depends on how risk averse they are. As long as the students justify their choices here, you could accept a wide variety of answers. Make sure it is clear to the students that the focus should be on the justification.

Activity 7 **Stock market trends and volatility**

Students will calculate the percentage gain or loss that could be made on the stock market by looking at a different starting point (purchasing price).

The purpose of selecting a new starting point to enter the stock market is to enable students to see that when you enter the market is as important as when you choose to exit. Based on this starting date, students will have made little money (and perhaps even lost money). This reinforces the idea that the higher the potential return, the higher the overall risk involved in the investment.

DP LINKS

Financial mathematics is studied in great detail in the mathematical studies course, and compound interest is revisited in standard and higher level.

Reflection

Yet again, it is important to emphasize that students must make valued justifications when prioritizing the factors to take into account when investing money. The order is not as important as the justification behind the ranking.

Summary

When exploring the concept of change—one of the fundamental and most important concepts in mathematics—it is also worth reflecting on how mathematics as a science has evolved and changed itself to meet the changes in the world and people—mathematicians and non-mathematicians. Traditionally, mathematics used to be divided into two types—applied and pure. However, in the last decades this border has faded as applied sciences have led to the invention of new mathematics: some pure mathematics that had been studied by passionate mathematicians solely for their own delight turned out to be very useful. (This is the case with number theory and its applications to secret codes, or abstract group theory and its application to crystallography.) Mathematics has also become a huge area of knowledge, so huge that no human being can learn all of it and it is still growing faster than ever before. Large amounts of mathematics are discovered or invented every year—new ideas, new theories and new applications appear constantly in scientific articles. Change is not only a concept *studied* in mathematics; it represents what is occurring *in* mathematics.

Equivalence

The state of being identically equal or interchangeable, applied to statements, quantities or expressions

	ATL skills	Mathematics skills
TOPIC 1 Equivalence, equality and congruence		
Activity 1 Big Mac index	✓ Negotiate ideas and knowledge with peers and teachers.	**Number** ✓ Convert between equivalent fractions. ✓ Use exchange rates to calculate values between currencies.
Activity 2 Medial triangles and congruency	✓ Make inferences and draw conclusions.	**Geometry and trigonometry** ✓ Determine triangle congruence by proving conjectures.
Activity 3 Congruent triangles in structures	✓ Compare conceptual understanding across multiple subject groups and disciplines.	**Geometry and trigonometry** ✓ Use Pythagoras' theorem, the cosine rule and volume formulas to solve real-life problems.
TOPIC 2 Equivalent expressions and equivalent equations		
Activity 4 Solving equations	✓ Recognize unstated assumptions and bias.	**Algebra** ✓ Rearrange equations or solve them for a variable.
TOPIC 3 Equivalent methods and forms		
Activity 5 Buying a car **Activity 6** A balanced diet	✓ Collect and analyse data to identify solutions and make informed decisions.	**Algebra** ✓ Determine linear equations to represent real-life scenarios. ✓ Find point(s) of intersection of linear systems of equations.
Activity 7 Equivalent forms of quadratic functions	✓ Evaluate evidence and arguments.	**Algebra** ✓ Compare different forms of a quadratic function to determine the best to use in a given context. ✓ Determine the key points of a quadratic function. ✓ Complete the square to derive the vertex form of a quadratic equation.
Activity 8 Will it make it?	✓ Apply skills and knowledge in unfamiliar situations.	**Algebra** ✓ Graph, factorize and solve quadratic equations. ✓ Determine a quadratic function, given a set of data points.
OTHER RELATED CONCEPTS	Representation Justification Simplification System	

Introducing equivalence

The concept of equivalence is such a subtle idea that students have been using it in mathematics for many years without even realizing it. To many, equivalence is simply being equal. However, in mathematics, this is not the case. Equality is just a simple example of equivalence and so this chapter aims to help students develop that understanding. They will be asked very often to reflect on their ideas about the concept of equivalence as they explore it in different contexts, formal and informal, abstract or connected to real-life problems. As they work through the activities they will realize that equivalence is a concept that has many facets.

In the introduction to the chapter, students are reminded of their first experiences working with equivalence: numerical expressions with the same value (numerical equality) and equivalent (or congruent) shapes. This should give students the idea that equivalence relates to something being equal in some way. However to challenge that notion, students are then reminded of the definition of equivalent sets (sets with the same cardinality or same number of elements). This is the beginning of their questioning of equivalence as a form of equality.

In Topic 1, students look at how equivalence relates to both numerical equality and congruence through applications of financial mathematics and geometry.

In Topic 2, students encounter activities that ask them to reflect about the meaning of equivalence in algebra: equivalent equations and identities. They also explore and think about the advantages of replacing expressions by equivalent ones when solving problems.

Finally, in Topic 3, the tasks challenge students to use equivalent methods and forms to solve problems and to reflect on which methods and forms might be more useful in given contexts.

> *Things that are equal to the same thing are equal to each other.*
>
> Euclid's first common notion

TEACHING IDEA 1:
Definitions

In case your students are not familiar with the definitions of equivalent and equal sets you may want to introduce them here. In fact, the topic of sets is an excellent context in which to introduce the concept of equivalence versus equality and to explore methods of proof or justification.

Start by giving students simple examples of equivalent sets:

$A = \{1, 2, 3, 4\}$ $B = \{2, 3, 5, 7\}$
$C = \{\text{prime numbers less than } 8\}$

and asking them to identify, with reasons, the sets that are equal. Then you can explore further the concept of equality of sets and discuss how to show that, in general, the identity
$A \cup (B \cap C) = (A \cup B) \cap (A \cup C)$ holds.

You can start by using Venn diagrams to represent both sets $A \cup (B \cap C)$ and $(A \cup B) \cap (A \cup C)$, and then discuss whether or not a diagram is enough to establish the truth of this identity. Ask: "Is a Venn Diagram acceptable as a proof of the statement?" This can then be linked to Chapter 8 on Justification, in which methods of proof are discussed.

TOPIC 1 — Equivalence, equality and congruence

Two numerical expressions are equivalent when they have the same value: they represent the same number. For example, the fractions $\frac{3}{6}$ and $\frac{1}{2}$ are equivalent; the expressions $2 + 3$ and $1 + 4$ are also equivalent. For these simple numerical cases, "equivalent" simply means that they have equal value. The symbol $=$ is used to express the equivalence between the numerical expressions. So, usually you write $\frac{3}{6} = \frac{1}{2}$ and $2 + 3 = 1 + 4$. Mathematical numerical equality is a permanent relation.

In the Big Mac index activity, however, currencies fluctuate and what is equal now may not be equal in a few days. As students determine whether or not a Big Mac has the same value in two currencies (once the exchange rate is taken into consideration) they should be made to realize that the result may be completely different at any other time. An equivalent amount of money in two countries may not be a permanent relation! So students need to watch for these nuances of the idea of "same value".

Activity 1 Big Mac index

In this activity students are asked to convert a monetary value from one currency to another. They will use current exchange rates and equivalent fractions. They will then use the Big Mac index to determine if one currency is overvalued or undervalued, compared to another, and hence predict whether that currency will depreciate or appreciate over time.

To ensure students can convert between different currencies, give them a simple budgeting task and let them create a holiday itinerary as follows.

You are going on a vacation abroad for one week and have (a set amount determined by the teacher in their home currency) to spend on your trip. You must find accommodation within your budget for the week, showing the total costs in both currencies. Be sure to include all calculations, including the current exchange rates that you used.

This task can be as simple or elaborate as you wish it to be. Students could simply choose an all-inclusive resort so they do not have to research and budget for meals. For a more elaborate task ask them to include meals and the costs of day trips and entrances to monuments, and so on. To be even more realistic, ask them to research the buy and sell price of currencies, so at the end of their trip they can convert their money back into their home currency.

Once students have explored the idea of equality or the same value in a financial application, they move into an exploration of what it means to be the same, or congruent, in geometry.

 Activities 2 and 3 — **Medial triangles and congruency**
Congruent triangles in structures

Both activities in this topic need a good understanding of congruency in order for the students to complete them successfully. Students should be familiar with how to determine if triangles are congruent by testing and proving conjectures. They will also need to know Pythagoras' theorem, the cosine law and the formulas for the volume of several appropriate three-dimensional shapes, including a pyramid.

TEACHING IDEA 2: Proving congruence

INQUIRY To prove congruence between two triangles, you look at three elements (any combination of sides and angles). Based on this idea, there are six possible combinations but only four prove congruence. The best way for students really to understand all of these combinations, and which four are the triangle congruence criteria (not including the right-angled triangle as a special case), is to test all possible combinations and see which ones hold true. If they can construct two different triangles with the same elements, then that combination of elements does not prove congruence. The congruence criteria can be established by taking one of them as an axiom and the others as theorems.

◯◯ WEB LINKS

While this investigation could be done with tracing paper or sticky notes, there is a good activity and applet that students can use to discover the theorems that will validate their results. Go to the Illuminations website, http://illuminations.nctm.org and search for "Congruence Theorems" and follow the detailed instructions.

EXTENSION

Let students create a piece of art or convert an existing famous piece of art into more of an abstract piece, using only congruent triangles and other congruent polygons.

 DP LINKS

SSA is not a congruence criterion and this is explored in DP mathematics when the ambiguous case of sine rule is studied.

Equivalent expressions and equivalent equations

Students should feel very comfortable with the procedures for solving linear equations, since they have typically been solving them for several years. However, they may not be as familiar with the justification for each of the steps. Understanding the reasons behind them will lead to a deeper understanding of why different types of equation may need to be solved in different ways.

👤 Activity 4 Solving equations

An expression is any mathematical string that represents a value, so, 1 would be a trivial expression. Then, for example, $x + 2x$ and $3x$ are equivalent expressions because, no matter which real-number value you assign to x, the numerical value obtained in both cases is the same. Although you would generally write $x + 2x = 3x$ for simplicity, to be rigorous you should actually write $x + 2x \equiv 3x$ to express this identity. An identity is a statement that two expressions are always equal in value, regardless of the values of the variables included in them. So if you have two such equivalent expressions, you can always use one of them instead of the other as they both produce the same numerical value for the same values of the variable. For this reason, identities are very important mathematical tools, used in simplification processes.

For more complex expressions you may need to consider only some values of the variable, since the expressions will only be valid in a certain set or domain. In this case, the identity symbol tells you to consider the equivalence between expressions in their common domain (that is, the intersection of their domains). For example, $\dfrac{x^2 - 1}{x - 1} \equiv x + 1$ only when $x \neq 1$.

Why does anybody even need equivalent expressions and identities? It is because they are useful when you need to simplify expressions and solve equations and inequalities.

As an introduction to this activity, you could ask students if there are general rules for simplifying expressions. In fact, there are very clear rules determined by the rules of the number system. For example, $2x + 3x = 5x$ (using the equality symbol informally) because $2x + 3x = (2 + 3)x$ by the distributive property of multiplication of real numbers with respect to addition. As this property is valid for all real numbers, you can use it when you work with expressions that include variables without any domain restrictions.

Another example is $\dfrac{2}{x/y} = \dfrac{2y}{x}$ when $x \neq 0, y \neq 0$. You could ask students to justify the equivalence between these expressions in the given domain.

Before students start this activity, they will need to understand how to apply the addition, substitution and multiplication principles when solving specific types of equation. Work through some basic examples, using each of the principles, so that students can see the process of simplification through equivalent expressions and ultimately find a solution.

💡 TEACHING IDEA 3: Equivalent expressions or identities

Set the students this short activity.

Decide whether the expressions in each pair are equivalent.

Give reasons.

If you need to impose domain restrictions, state them.

a) $2x + 3x + 5x$ and $10x$

b) $2x + 3x - 5x$ and 0

c) $x + \dfrac{x}{2}$ and $\dfrac{3x}{2}$

d) $\dfrac{3}{2}x$ and $\dfrac{3x}{2}$

e) $\dfrac{x^2 - 9}{x - 3}$ and $x + 3$

f) $\dfrac{1 - \sin^2 x}{\cos x}$ and $\cos x$

g) $\sqrt{x^2}$ and x

At this stage you could use this activity to include the zero product law.

TEACHING IDEA 4:
More complex equations

Set questions such as these, to provide practice.

Solve each of these equations.

For each step, decide whether to move to an equivalent equation or if you may need to restrict and watch for extraneous solutions.

a) $\dfrac{4}{x-1} = \dfrac{5}{(x-1)^2}$

b) $\sqrt{x^2} = 2\sqrt{x}$

c) $\sqrt{x-6} = x$

The three principles may not be enough when you need to solve more complex equations. You will often need another very important law.

Zero (or null) product law: If the product of two expressions is zero, then at least one of the expressions must be zero.

This law provides a very efficient method for solving complex equations. It converts the problem of finding the solution of a complex equation into a collection of simpler equations.

For example:
$x^2 - 4 = 0 \Rightarrow (x-2)(x+2) = 0 \Leftrightarrow x - 2 = 0 \text{ or } x + 2 = 0$
The original equation is therefore equivalent to two simpler equations. The union of their solution sets gives the solutions of the original equation.
So $x^2 - 4 = 0$ has two solutions: $x = \pm 2$.

A more complex example is $\dfrac{2}{x} = \dfrac{3}{x^2} \Rightarrow \dfrac{x}{2} = \dfrac{x^2}{3}$, where $x \neq 0$. In this case

the first equation is not equivalent to the second because the first has a domain restriction. However, the first equation implies the second, because every non-zero real number has a unique reciprocal. So, as you continue solving the equation, you need to remember that you are working with non-zero real numbers.

$\dfrac{2}{x} = \dfrac{3}{x^2} \Rightarrow \dfrac{x}{2} = \dfrac{x^2}{3} \Leftrightarrow 3x = 2x^2 \Leftrightarrow 3x - 2x^2 = 0 \Leftrightarrow x(3-2x) = 0 \Leftrightarrow x = \dfrac{3}{2}$

Note that in the last step $x = 0$ is not a solution because zero is not in the domain of the original equation. Although $x = 0$ appears as a solution of subsequent equations, it is in fact an extraneous solution that you must discard.

Equivalent methods and forms

Problems in mathematics can often be approached in variety of ways. In this topic, students are first asked to look at the different methods that can be used to solve systems of linear equations and then to reflect on the advantages and disadvantages of each.

Activity 5 provides a real-life context for a 2-by-2 system of equations. Activity 6 challenges students to solve an application of a 3-by-3 system. The remainder of the topic is spent investigating the different forms of a quadratic function and then applying them to the analysis of a ball being kicked in sports.

Throughout this topic, students are asked to reflect on how the current activity relates to previous ones in the chapter. While equivalence can be a difficult concept to define succinctly, students can see how its different aspects are interrelated.

Equivalent methods

While equivalence sometimes refers to equality, it can also be explored in terms of the different methods used to solve a problem that result in the same answer. In Activity 5, students look at how long it will take for two cars to cost the same, once the initial price and gas consumption are taken into account.

Activity 6 asks students to use a 3-by-3 system of equations to analyze their diet. In both of these activities, students use equivalent methods to solve the problem and then reflect on which method they prefer.

Systems of equations: 2-by-2 system of equations

 Activity 5 Buying a car

Assessment

This activity could be a summative task at the end of a unit on systems of equations.

If you choose to assess students on this task, you can use criterion D. The task-specific descriptor in the top band (7–8) should read that the student is able to:

- identify all factors and assumptions that need to be considered for this task
- determine the correct linear equations for both cars
- correctly interpret the gradient and intercept on the vertical axis of both linear equations
- correctly graph the linear system
- correctly determine the point of intersection of the linear system, using two methods
- correctly interpret the point of intersection of the linear system, given the context of the problem
- justify the degree of accuracy used in all calculations
- critically evaluate both methods for solving the linear system and justify which method is the more appropriate to use given the context of the problem.

TIP

The reflection questions in the task are essential if students are truly to address the statement of inquiry so that they critically analyse consumer choices and take environmental issues into consideration.

Key concept	Related concepts	Global context
Relationships	Equivalence Model	Globalization and sustainability
Statement of inquiry		
Modelling relationships to represent real-life scenarios allows us to compare different options in order to minimize our impact on the environment.		

TEACHING IDEA 5: The number of solutions to a linear system

INQUIRY Students could perform an investigation to determine the number of possible solutions (points of intersection) of a linear system of equations.

Present them with a copy of the table below, showing three linear systems. Ask them to graph each pair on a separate set of axes. As they graph each system, they can then fill in the last column.

System	First equation	Second equation	Number of solutions
A	$y = 2x + 1$	$y = -x + 7$	
B	$2y + 8 = x$	$4y - 2x = -16$	
C	$6x = 20 - 2y$	$y + 3x - 5 = 0$	

Ask students whether they think it is possible to determine how the lines will intersect before actually trying to find the solution by graphing. After a brief discussion, let them fill in their tables.

	A	B	C
Linear system in $y = mx + b$ form			
Gradient of the line			
y-intercept			
Number of solutions			

Ask them to write a brief summary of their investigation of the three types of linear system.

Finally, ask students to solve the three linear systems, using algebra, and then compare their answers to the ones obtained previously. What do they notice? Ask them to explain how the algebraic and graphical answers are related.

REFLECTION

Ask students to give a real-life example of when each of the three types of linear systems could occur.

TIP

You can use this activity as an introduction to linear systems, for students to see the three different cases, but it is too guided if you want to assess their work against criterion B. To make it appropriate for assessment with criterion B, you should remove the scaffolding so that students can select their own problem-solving methods, create their own tables to see patterns, make generalizations and conduct some sort of test to justify their generalizations. Leaving these steps up to students will ensure that they have the ability to reach the highest achievement level.

Systems of equations: 3-by-3 system of equations

 Activity 6 — **A balanced diet**

Assessment

In this activity, students apply their knowledge of 3 by 3 systems of equations to the analysis of their diet. This is an appropriate summative task for a unit on systems of equations.

If you choose to assess students on this task, you can use criterion D. The task-specific descriptor in the top band (7–8) should read that the student is able to:

- identify all factors and assumptions that need to be considered for this task
- correctly determine the variables
- determine the correct equations for protein, fat and carbohydrate (one each)
- correctly simplify the system of equations to a 2-by-2 system
- correctly solve the resulting 2-by-2 system of equations
- correctly interpret the solution in the context of the problem
- justify the degree of accuracy used in all calculations
- justify whether or not he/she would want to meet the daily requirements with the chosen foods.

Stage 1 of the unit planner

Key concept	Related concepts	Global context
Logic	Equivalence Representation Simplification	Identities and relationships
Statement of inquiry		
A logical process of simplification can help us achieve a healthy lifestyle.		

It is very common for students to solve the system but find that one or more answers represents a negative number of servings. In this instance, this should be interpreted as "no real solution". This does not mean that the system is "inconsistent" since it, indeed, has a solution. It is just that the solution does not make sense in the context of the problem.

> **TIP**
>
> Students should try to be very specific when typing in foods on www.dietfacts.com. It offers many choices of "pasta" or "yogurt" which can slow down the process. Encourage students to round quantities to the nearest unit in order for the system to be more manageable.

EXTENSION

Students can solve 3-by-3 systems in a variety of ways. While "elimination" is probably the most common, students can also use substitution or be taught how to use their GDC to solve a system of equations. You could teach more advanced students row operations on matrices (often referred to as Gaussian elimination), as another way of representing the elimination method.

TEACHING IDEA 6: Analyse your diet

Rather than asking students simply to solve a random system of equations in class, let them represent a real-life situation instead. Perhaps use foods that you eat and attempt to balance other nutritional elements, such as iron, fibre, vitamin A, sodium, and so on. Also, the recommended daily allowance (RDA) is for an entire day, whereas the foods in the table are generally for one meal. Adjusting the RDA amounts to reflect the size of the meal may make finding a workable solution easier.

Equivalent forms

It is important for students to realize that, in mathematics, they often have choices. As seen in the previous activities, problems may be solved in a variety of equivalent ways. Similarly, many functions have a variety of representations from which students can select. In the remaining activities, students will investigate equivalent forms of a quadratic function and how they can be applied to a real-life situation.

 Activity 7 **Equivalent forms of quadratic functions**

The different forms of quadratic functions are a focus in the final years of the MYP. Students must understand how the different forms relate to each other and how they can be used to solve problems efficiently. In this activity, students are investigating all of the key points of the quadratic function and the advantages and disadvantages of each form.

After completing all three steps, students should summarize their findings in a table like the one below. Let them either put a tick [✓] in those columns where the information is readily available from the given form or rate the difficulty of finding the information [for example, "easy", "medium" or "hard"].

	Standard form	Factored form	Vertex form
x-intercepts			
y-intercepts			
Axis of symmetry			
Vertex			

To consolidate their understanding of the charts they have created, ask them to answer the following questions.

Which form of the quadratic function is most useful when you want to investigate:

a) the graph showing the function's vertex

b) if the function has a minimum or a maximum

c) the point of intersection of the graph of the function with the x-axis

d) the point of intersection of the graph of the function with the y-axis

e) points of intersection with other functions

f) the intervals where the function's domain values are positive and where they are negative

g) the intervals where the function is increasing and where it is decreasing.

CHAPTER LINKS

In Chapter 2 on the key concept of Form there are many tasks relating to the different forms of quadratic functions that could be used to complement the tasks in this topic.

TIP

Students should already know the quadratic formula and how to use it, before doing this inquiry, so they see the connection between forms.

Discussion

Ask students for their opinion on which of the three forms is the easiest to graph. Encourage them to use an example to explain their choice.

Ask: "Would it be possible to solve all quadratic questions knowing only the standard form? Explain why or why not."

TEACHING IDEA 7: The quadratic formula

INQUIRY Write up the standard form of a quadratic equation $ax^2 + bx + c = 0$, $a \neq 0$. Ask students to isolate x by completing the square. Ask: "What is your final answer?" Explain this connection to the concept of equivalence between forms of quadratic functions. They could also relate this to identities and the substitution principle explored in Topic 2.

Game—Creating a loop

An interesting way to test/consolidate student knowledge of the three forms is by creating a set of cards that each have two separate and distinct equations in different forms on them. Students will receive one card and will have to convert both equations into the other two forms. There will be one other person in the class who has an equivalent form to each of their equations. When they find that person they link up and continue to try to find their missing equation pairs. The game ends when the last equation has been matched and a closed loop has been formed.

As a variation on the game above, create cards that each have one equation on them. Students have to convert this equation into the other forms (and graph it if needed) and then find their equivalent equations (and graph). You could require them just to find one other person, or two, or three by including a set of cards with the graph on it too.

WEB LINKS
http://illuminations.nctm.org has a great visual on the act of completing the square and visually explains the process. Search for "completing the square" in the "interactives" section.

TIP

This activity is also an effective way to group students randomly for group work—anything from pairs to groups of four.

Activity 8 Will it make it?

In this activity, students will use different forms of quadratic functions to solve a problem in a real-life context. Given a set of data points, they will determine a quadratic function that models the data so that they can make a prediction.

Before they complete this activity, let students watch the actual video of the kick, which can be found on www.youtube.com. Search for "David Akers 63" and then show the video, stopping it just before the result is known. Let students perform the activity and then find out the "real answer" by watching the end of the video.

Once students understand the advantages and disadvantages of each form, and when to use each of the forms given the context, they can move on to solving real-life problems. The football task requires students to answer a variety of questions, using the most appropriate method. One of the questions asks what assumptions have been made, which may be difficult for the students to answer. They have assumed that the football follows a perfect quadratic equation without testing if this is the case. You could ask the students to conduct a quadratic regression to confirm that all of the given points fall on the graph.

INTERDISCIPLINARY LINKS
Students could perform a similar activity with data they collect from kicks performed during their physical and health education class or even with one of the school's sports teams.

TEACHING IDEA 8: Exploring quadratic equations

You could use this activity after students have completed the football task. It is a good example of how most real-life applications do not have numbers that will complete the square or factorize easily. Students need strategies for dealing with these sorts of question and must gain confidence in using the quadratic formula.

Instructions for students

The height of a projectile that is tossed vertically follows the same quadratic function $H(t) = h_0 + Vt - \frac{1}{2}gt^2$, where h_0 is the initial height of the projectile in metres, v_o is the initial speed in metres per second (m/s), and g is the acceleration of gravity in metres per second squared (m/s²).

What form of the quadratic equation does this represent? Why do you think the projectile motion is expressed in this form?

This formula can actually be used on any astronomical body provided that you know the acceleration due to gravity of that body. On Earth, the acceleration of gravity is approximately $g = 9.8$ m/s^{-2} and on the Moon it is approximately $g = 1.6$ m/s².

When solving these problems, explain what key points you need and what form of the equation you worked with to solve them (if you altered it into a different form).

a) Assuming the initial velocity of the projectile being thrown is 15 metres per second from an initial height of 1.5 metres, how much higher will the projectile reach on the Moon compared to Earth?

b) How much further will it travel once it hits the ground?

c) The acceleration due to gravity on Earth is approximately six times that on the Moon. Does that mean that the projectile will go six times higher and further on the Moon than on Earth? Explain.

⊂⊃ DP LINKS

Projectile motion is an important topic in DP physics courses. Students often need to solve quadratic equations in their study of the motion of objects, though they are usually given in standard form.

Summary

Equivalence is a concept that seems relatively simple but, in fact, has many different facets. Students began the chapter by returning to the idea of equivalence in terms of equality and congruence. They then extended their understanding of the concept by looking at how equivalent equations are produced by the application of fundamental principles while solving equations. Finally, students explored how equivalent forms and methods can be used to represent mathematics and solve problems. Equivalence is much more than simply being equal, something students should now be able to understand and explain.

Generalization

A general statement made on the basis of specific examples

	ATL skills	Mathematics skills
TOPIC 1 Number investigations		
Activity 1 Climbing stairs	✓ Use appropriate strategies for organizing complex information.	**Number** ✓ Discover and make generalizations about number patterns. ✓ Use recurrence relations to determine terms of numerical sequences.
Activity 2 Bad apples–it takes only one to spoil the crate!	✓ Use appropriate strategies for organizing complex information.	**Algebra** ✓ Analyse and use well-defined procedures for solving complex problems.
TOPIC 2 Diagrams, terminology and notation		
Activity 3 Can he do it?	✓ Use models and simulations to explore complex systems and issues.	**Geometry and trigonometry** ✓ Construct diagrams and use appropriate notation to explore networks to discover patterns and rules associated with them.
Activity 4 The pony club problem	✓ Use models and simulations to explore complex systems and issues.	**Geometry and trigonometry** ✓ Construct diagrams and use appropriate notation to explore networks to discover patterns and rules associated with them. ✓ Construct diagrams to explore the medial triangle to discover its properties.
Activity 5 The medial triangle	✓ Test generalizations and conclusions.	**Geometry and trigonometry** ✓ Construct diagrams to explore the medial triangle to discover its properties.
Activity 6 Exercising caution	✓ Practise observing carefully in order to recognize problems.	**Number** ✓ Discover and make generalizations about number patterns. **Algebra** ✓ Use recurrence relations to determine terms of numerical sequences.

Activity 7 Vertical stretches	✓ Draw reasonable conclusions and generalizations.	**Algebra** ✓ Investigate the basic transformations of rational functions.
Activity 8 Reflections		
Activity 9 Translations		
Activity 10 More translations		
Activity 11 Summary of generalizations		
Activity 12 Further examples to validate your general rules		
OTHER RELATED CONCEPTS	Pattern Representation Justification Space	

Introducing generalization

In this chapter, students are challenged to expand their problem-solving skills by using a collection of explicit strategies originally developed by the Hungarian mathematician and educator, George Pólya, during the last century. Pólya's ideas are based on the assumption that learning mathematics means learning how to think, how to discover rules and also how to invent methods. According to Pólya, it is all about how to handle abstractions. For example, to solve a practical problem, you must first make an abstract problem. You need to translate the information into equations, or define functions that model the situation. Ultimately, to understand mathematics means to be able to do mathematics and this means being able to solve mathematical problems.

At the start of the chapter in the student book, there is a comprehensive list of problem-solving strategies. Throughout the chapter, students have a diversity of tasks that give them the opportunity to use these strategies. However, this list is a particularly useful resource to be used often in class or at home, whenever students solve problems in context. The list is a collection of tips and hints to help students think mathematically.

> *Teaching must be active, or rather active learning... the main point in mathematics teaching is to develop the tactics of problem solving. What is good education? Systematically giving opportunity to the student to discover things by himself.*
>
> George Pólya

WEB LINKS
Visit www.youtube.com and search for "problem-solving with George Pólya"

TOPIC 1 Number investigations

Number is the first and most common abstraction students need to deal with in school mathematics. The activities in Topic 1 are about number patterns.

 Activity 1 **Climbing stairs**

In this activity, students rediscover a well-known number pattern: the Fibonacci sequence (1, 1, 2, 3, 5, 8, …) and its recurrence definition (start with 1, 1 and then, to obtain the next term, add the two previous ones).

Students need not have previous knowledge of this sequence but if they do it is easier to spot the pattern. As they complete the activity, they will recognize the importance of using simple notation and organizing data systematically, to help in analysing it and identifying patterns.

They will discover the Fibonacci sequence and generalize why the rule works, in the context of the task. Throughout the activity, students will apply a variety of problem-solving techniques, including the use of diagrams or tables to display information.

EXTENSION

Fibonacci numbers have interesting arithmetic properties that are worth exploring. For example, the extension activity offers an opportunity to explore further problem-solving techniques and leads the students to generalize a property of consecutive Fibonacci numbers.

Take any four consecutive Fibonacci numbers, such as 1, 1, 2 and 3.

Use them to find the values of a, b and c as follows .

To determine the value of a, multiply the two middle numbers and double the result. (Here, multiply 1 and 2 to give 2, then double 2 to get 4, so $a = 4$.)

The product of the two outer numbers, gives the value of b. (Here, 1×3 is 3, so $b = 3$.)

To get the value of c, add the squares of the two middle numbers. (Here, $1^2 + 2^2 = 5$, so $c = 5$.)

Try other sequences of four consecutive Fibonacci numbers and complete the table.

Fibonacci numbers	a	b	c	$a^2 + b^2$	c^2
1, 1, 2, 3	4	3	5		
2, 3, 5, 8					

What do you observe? Write down your conjecture, test it further and write down your conclusion.

The table will make clear that $a^2 + b^2 = c^2$.

TEACHING IDEA 1: Climb in pairs

Students could try this activity for themselves and climb stairs according to the rules of the task. They could do this in pairs. One student tries all different ways of getting to step 1, 2, 3, … and the other student records the number of ways this can be done.

TEACHING IDEA 2: Extending the activity

Extend the suggested activity and explore it for any Fibonacci type sequence, such as 2, 4, 6, 10, 16, … and generalize the result obtained. You can take this activity further and actually use algebra to prove the result. If the first two terms are x and y then the four consecutive terms are $x, y, x+y, x+2y$ and therefore $a = 2xy + 2y^2$, $b = x^2 + 2xy$ and $c = y^2 + (x+y)^2$. Expanding the squares and simplifying, it is fairly easy to show that $a^2 + b^2 = c^2$ (both sides are given by $x^4 + 4x^3y + 8x^2y^2 + 8xy^3 + 4y^4$).

Another interesting property of a Fibonacci-type sequence is that the sum of 10 consecutive terms is equal to 11 times the 7th term. This can be a rewarding challenge for students with good algebraic skills and is a nice extension to the activity here.

TEACHING IDEA 3: Fibonacci and Fibonacci numbers

 INQUIRY Leonardo de Pisa (1175–1250), or Fibonacci, is mostly known because of the famous sequence that holds his name and that is simply the solution to a problem that he included in his book *Liber Abaci*.

However, the main contribution to mathematics made by Leonardo de Pisa was the introduction of the Hindu-Arabic number system in Europe. The study of Fibonacci numbers offers a good opportunity for students to learn more about the sequence and its applications but also about the mathematician. They may be impressed to learn how his work changed the course of mathematics history by making calculations easy and accessible to many people.

Students could work in groups to research different aspects:

- Fibonacci numbers in nature
- Fibonacci numbers and the golden ratio
- Curious properties of Fibonacci numbers (for example, its relation with Pascal's triangle)
- Fibonacci the mathematician and the Hindu-Arabic system.

Activity 2 — Bad apples—it takes only one to spoil the crate!

Assessment

If you choose to assess students on this task, you can use criterion D. The task-specific descriptor in the top band (7–8) should read that the student is able to:

- identify the real-life elements of the apple-packing problem
- select appropriate mathematical strategies, such as constructing appropriate diagrams and tables, at the various stages of the problem
- apply the mathematical strategies to arrive at a correct conjecture for all of the various crate sizes and dimensions
- justify their algebraic formulae by testing them within the scope of the problem
- justify whether their solutions make sense in the real-life situation that is depicted.

Stage 1 of the unit planner

Key concept	Related concepts	Global context
Relationships	Generalization Space	Scientific and technical innovation
Statement of inquiry		
Understanding naturally occurring interactions and generalizing these relationships can help us minimize resource wastage.		

In this activity, students will use some of Pólya's problem-solving techniques, in particular making a plan, simplifying a problem, and using diagrams and tables. It would be pedagogically valuable to introduce this task by starting out with the final question, that is, how many days would it take a rectangular box of apples that are packed in a particular way to rot, given that the box initially contains one rotten apple? (Students should realize that many other factors would have to be considered to simulate this in a real-world context. They will do this at the end, as part of their reflection, but for the activity itself, they will start with simplified initial conditions.)

In order to solve this problem, a Pólya-style plan has been created for them, which begins with an easier version of this problem. They are asked to find the time needed for all the apples in one layer packed in a square box to rot. Students should have the following results for Step 1.

Size of tray	Spoiled apple	After 1 day	After 2 days	After 3 days	After 4 days	After 5 days
4 × 4	5	1, 6, 9	2, 7, 10, 13	3, 8, 11, 14	4, 12, 15	16

The same total amount of time, 5 days, will be needed if the starting apples are: 2, 3, 8, 9, 12, 14, 15.

Size of tray	Spoiled apple	After 1 day	After 2 days	After 3 days	After 4 days	After 5 days	After 6 days
4 × 4	1	2, 5	3, 6, 9	4, 7, 10, 13	8, 11, 14	12, 15	16

The same total amount of time, 6 days, will be needed if the starting apples are 4, 13, 16.

Size of tray	Spoiled apple	After 1 day	After 2 days	After 3 days	After 4 days
4 × 4	6	2, 5, 7, 10	1, 3, 8, 9, 11, 14	4, 12, 13, 15	16

The same total amount of time, 4 days, will be needed if the starting apples are 7, 10, 11.

Conclusion

If the first bad apple is at a corner, it takes the longest for the entire tray to go bad, six days. If the first bad apple is on a side but not in the corner, it takes five days for the tray to go bad, and if the first bad apple is anywhere in the middle, then the tray will go bad in four days.

Conjectures and reasons

For any n by n square grid of apples, take D as the number of days for all apples to go bad.

1. The longest time for the tray to go bad will occur when the first apple to go bad is at a corner. This may best be explained by counting the boundaries that need to be crossed, rather than the apples themselves. A boundary is the point of contact between two adjacent apples. If an apple is in the top left corner of an n by n box, then there are $(n-1)$ boundaries between it and the top right corner, and $(n-1)$ boundaries between that and the bottom right corner. Therefore there are $(n-1) + (n-1) = (2n-2)$ boundaries to traverse before the last apple goes bad. Hence, the formula is: $D = 2n - 2$.
2. The shortest time for the tray to go bad when the first apple to go bad is not a corner or side apple, i.e., anywhere in the middle, is $D = 2n - 4$.
3. The middle time for the tray to go bad is when the first apple to go bad is along the side but not anywhere in the middle position. A formula is $D = 2n - 3$.

Students should confirm their formulas by using other square grid sizes. They should conclude that, for any square grid size, the formulas above hold.

They use the same methodology and reasoning to investigate rectangular grids. They should obtain that, for an m by n grid:

■ if the first bad apple is at a corner, then $D = (m + n) - 2$
■ if the first bad apple is on the side, then $D = (m + n) - 3$
■ if the first bad apple is anywhere in the middle of the grid, then $D = (m + n) - 4$.

Steps 4 and 5 are good tasks for group work. Students consider multiple square layers by bringing in the next condition, that apples can be contaminated not only if they are packed side by side, but also on the top and bottom. Eventually, they develop a formula for the number of days for all apples in a square crate to rot, which is then extended and generalized for any rectangular-shaped crate.

For apples in a 3D m by n by k crate to go bad, assuming that those underneath also just touch those above in one spot, you obtain the result by adding $(k-1)$ to the results obtained above. For example, if the first apple to go bad is in the corner of the top tray, then $D = 3n - 3$, and so on.

After doing this task, a teacher-led discussion on which Pólya strategies were useful in solving the problem would help consolidate some of the problem-solving strategies.

TOPIC 2 — Diagrams, terminology and notation

In this topic, students are asked to use diagrams to solve problems. They are introduced to specific terminology to express their conclusions concisely and correctly. The student book offers a task on discrete mathematics, with some possible extensions suggested in this section. Alternatively, the same ideas can be explored using dynamic geometry software. An alternative task and corresponding reflection can be found at the end of this section.

Activity 3 — Can he do it?

The postman and circuits problems

In this activity, students are guided to explore a famous problem in discrete mathematics: Euler graphs and circuits. Different versions of this problem are fairly accessible and require no prior knowledge of the topic. For this reason they are good investigational activities that can be adapted and used with different year groups to improve students' problem-solving skills.

Students are given a collection of diagrams and their task is to decide whether it is possible to start at a vertex (or node) and move along each edge (or arc) exactly once and maybe even return to the starting point.

Students should eventually make the following discoveries.

- It is possible to go around the graph and return to the same point, using each edge exactly once, if and only if all the vertices have an even degree (they meet an even number of edges).
- If there are exactly two vertices with an odd degree, it is still possible to go around the graph, using each edge only once, if these vertices are exactly the start and end points of the route.
- The sum of the degrees of all the vertices of a graph must be even.
- In every graph, the total number of vertices with odd degree is always even.

The second part of the activity introduces some graph terminology and the students need to answer the last parts of the questions.

TEACHING IDEA 4: Colouring maps

If students are working with graphs for the first time, you could introduce the topic with an activity that uses maps and colours. What is the minimum number of different colours that you will need to colour in a map if two sections that share a common edge cannot be coloured the same?

Let the students discuss their ideas, then give them a map (something with 7–10 states/countries on it as you want to keep it relatively simple) and ask them to devise a method for finding a solution. No doubt they will have some interesting ideas, but the general theme will be that they will have to produce a simplified diagram to help them solve the problem. This is a useful introduction to networks because you can show them how to create a simple network in which each state/country is represented by a vertex. Then colour each vertex, ensuring no two vertices that share an edge are the same colour. Students will discover that, at most, four different colours are needed to shade any map. This is called the four-colour theorem.

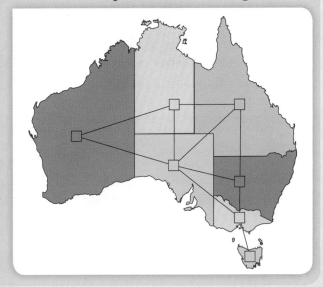

For example, the seven states of Australia (with Tasmania included even though it would not have to be as it is not connected to the mainland) can be represented by the simple network shown here.

Then let students choose a map (or geometric picture), create a simplified network to represent it, and colour it in using only four colours.

This activity will lead in well to Euler paths and circuits.

This activity provides an additional opportunity for students to use graph terminology and explore the advantages of using simple graphs to represent each situation. Once each course is represented by a simple graph, where each field is represented by a vertex and each gate by an edge, students realize quickly this activity is, in fact, identical to Activity 3. Both can be represented by a simple graph. In the first case the vertices represent the islands and in the second case they represent the fields. In the first case the edges represent the bridges and in the second case they represent the gates. Therefore, the general result is the same. Moreover, the same result can be generalized and extended to any situation represented by a simple graph, that is, the Eulerian circuit exists if and only if all the vertices are of even degree.

Depending on the year group you are working with, it might be necessary to scaffold the task to ensure the students are looking for the right elements within this problem. If necessary, you could tell them to set up a table with these columns.

	Number of even vertices	Number of odd vertices	Euler path? (yes or no)	If yes—start and finish (at which type of vertex?)
Example 1				

TEACHING IDEA 5: Plan a trip
Plan a trip to an amusement park of your choice. Research information about the attractions you would like to visit and rides you would like to try. Draw a graph in which the vertices represent the attractions (choose a labelling system that is easy to read) and the edges are labelled with the time it takes to get from one attraction to the next. Plan the visit so that you have time to do all the activities that you want to. What is the minimum time needed? Remember to research the average waiting time for each attraction. In which order should you visit the attractions?

TEACHING IDEA 6: The travelling salesman problem
Research this famous problem. Create a simple example where the salesman has to visit five cities of your choice. Draw a graph showing the cities and label the edges with the distances between them. Find the optimal route. This will be the shortest one that allows the salesman to visit all the cities. Search online for a simulator and explore the problem for an increasing number of cities to visit. Relate this problem to other similar ones: finding the shortest route between two points in a city, using public transport, or the cheapest ticket you can buy to travel from city A to city B, using any airline and allowing multiple stops. Reflect on the importance of this area of mathematics and on the limitations of the algorithms available.

TIP
The added complexity of having a weighted network (a network where edges have a certain "value" or "weight") is a useful extension of the basic Euler path. The task in teaching idea 6 is an open-ended problem that could be used as an assessment task because it has the potential to differentiate easily and all students can have success with the task and see it through to completion, even if they do not complete it as a weighted network.

WEB LINKS
The Demonstration Project from Wolfram Research offers interactive applets that make the exploration of teaching idea 6 easy and interesting. See http://demonstrations.wolfram.com

TEACHING IDEA 7: Planar graphs of solids

Research planar graphs of solids—what are they? How can they be used to prove Euler's formula for solids: $V-E+F=2$? How can this formula be used to show that there are exactly five regular solids (tetrahedron, cube, octahedron, dodecahedron and icosahedron)? What are dual solids? How can you use the graphs of the Platonic solids below to identify duality? Why are the regular solids called Platonic solids and why were they very important for the ancient Greeks?

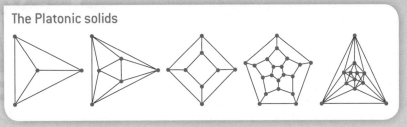

The Platonic solids

TIP

In the "Reflecting on diagrams, terminology and notation" section of Topic 2, students are asked to suggest questions about networks in their everyday lives. Collect these answers to check them, then distribute them as an excellent review for the unit.

All of these teaching ideas require students to deal with new concepts and explore a substantially different area of mathematics in which there are very few theorems to use, but rather, general strategies that need to be adapted to each situation. These are excellent opportunities to apply problem-solving strategies in context and develop resilience.

Activity 5 — The medial triangle

This activity offers the possibility for students to recognize the importance of the use of good diagrams, notation and terminology while discovering results from another branch of mathematics: geometry. To be able to complete this activity, students will have to know the definitions of a centroid, orthocentre and circumcentre and how to construct them. They will then discover the relationships between the properties of a triangle and its medial triangle and test these discoveries, to make generalizations.

In this activity, students explore the geometric properties of these special points of triangles. While using dynamic geometry software, students may discover some of the following properties about medial triangles.

- The perimeter of the medial triangle is one half that of the original triangle.
- The area of the medial triangle is one quarter that of the original triangle.
- The medial triangle divides the original triangle into four smaller, congruent triangles that are each similar to the original triangle.
- The centroid of a triangle is also the centroid of its medial triangle.
- The circumcentre of a triangle is the orthocentre of the medial triangle.

There are more conjectures about collinear points that can be researched if you want further extensions for your students.

TIP

Using dynamic geometry software can help students make many observations, see patterns and hence make generalizations about many geometric properties. In the suggested activities, students will make generalizations about the locations of the different triangle centres to help them determine which would be best to use in different real-life situations.

TEACHING IDEA 8: Euler's line and nine-point circles

INQUIRY These are some inquiry-based extension activities to be explored with a dynamic geometry program.

a) The Euler's line, named after Leonhard Euler, is the line passing through the orthocentre H, the circumcentre O and the centroid G of any triangle. Construct a sketch and verify that the orthocentre H, circumcentre O and centroid G lie on a straight line. Find the relative distances between the points H, O and G, then move a triangle vertex to look at all types of triangles. What do you notice about these distances?

b) The nine-point circle is a circle that can be constructed for any given triangle. It is so named because it passes through nine significant points:

- the midpoint of each side of the triangle
- the foot of each altitude
- the midpoint of the segment of each altitude from its vertex to the orthocentre.

The centre of any nine-point circle (the nine-point centre) lies on the corresponding triangle's Euler line, at the midpoint between that triangle's orthocentre and circumcentre. Construct a dynamic sketch showing the nine-point circle of a triangle of your choice. Drag the vertices of the triangle and test these properties of this circle.

1. The radius of any nine-point circle is half the length of the radius of the circumcircle of the corresponding triangle.

2. Any nine-point circle bisects any line from the corresponding triangle's orthocentre to a point on its circumcircle.

CHAPTER LINKS
The conjectures above can be justified using proofs—see Chapter 8 on Justification for strategies on how to conduct visual proofs.

CHAPTER LINKS
In Chapter 15 on Space, students are asked to construct a mobile structure using at least six different shapes after completing an investigation on the centroid. They could try Activity 5 before constructing the mobile. This will help to consolidate their understanding of triangle centres and confirm that the centroid is the best choice when creating the mobile, as it is always inside the triangle, regardless of the type of triangle used.

TIP

The important steps in the activity are the creation of the diagrams and construction of the points to help make the discoveries. Using dynamic software is essential so that all discoveries for all types of triangles can be tested easily and generalizations can be made.

Beware of quick generalizations!

The goal of this section is to make students wary of making generalizations too quickly, or not testing sufficient cases before making a generalization. A teacher-led discussion about the clever quotes in the introductory paragraph of this topic would be helpful. This section should be linked to the paragraph on justification; students should realize that before claiming a mathematical generalization to be true, they must subject it to the rigorous process of proof.

To demonstrate the caution that should be exercised in making generalizations, you could use the teaching idea below, as a group warm-up before students attempt this activity.

TEACHING IDEA 9: Chocolate squares

Students work in pairs. They work out how many moves it takes to break a chocolate bar into squares. They record the number of moves necessary to achieve single squares. They can use graph paper or you may choose to let them use actual chocolate bars.

Let students consider a typical chocolate bar that consists of m by n squares. They should find the minimum number of moves needed to break, for example, a 4 by 5 bar into individual single squares.

Students will discover that the number of moves needed to break an m by n chocolate bar, independent of the starting point or size of chocolate bar, is $mn - 1$. A discussion can follow on why this is the case.

Students are now ready to think about whether some generalizations have limitations, or whether they are true for all cases.

STEP 1 In step 1 of Activity 6, students look at an easy and famous number pattern. A consideration of how to multiply two or more digits by two or more digits easily yields the explanation for the pattern generated, as well as limitations on the pattern. They should explain these limitations carefully, and use properties of numbers and integer multiplication to justify them.

STEP 2 In step 2, students consider the relationship between the number of points on the circumference of a circle, and the number of regions made when the points are connected. From the given information, students will readily conjecture that if n is the number of points on the circumference, then there are 2^{n-1} regions. Testing the number of regions for 5 points on the circle does give 16 regions, but for 6 points it does not give 32, as the suggested rule implies, but 31 regions!

The actual conjecture and its proof is part of graph theory in discrete mathematics. The first part of this proof is easily accessible. They will readily see by experimentation that each intersection inside the circle is defined by the intersection of two lines, and each line contains two points so that every intersection within the circle defines four unique sets of points on the circle's circumference. This can be seen in the fourth circle on page 63 in the student book. Each vertex on the circle has $n-1$ edges, and each vertex inside the circle has four edges. Students will need help, however, in formulating that, hence, there are n vertices on the circumference, but $_nC_4$ or $\binom{n}{4}$ vertices inside the circle.

Notation: $_nC_r$ and $\binom{n}{r}$ **mean the number of combinations of n objects taken r at a time, so from n objects we make a selection of r of them.**

The number of lines, or edges, is therefore $\dfrac{n(n-1)+4\binom{n}{4}}{2}$ (since each edge would otherwise be counted twice by its two vertices).

Then, using Euler's formula, $V - E + F = 2$ (where V is the number of vertices, E the number of edges, and F the number of faces), and solving for F, you get $F = 2 + E - V$. However, the number of regions is the number of faces+1, since the region outside the circle is also counted. Hence, the number of regions is $1 + E - V$, which reduces to $_nC_4 + _nC_2 + 1$.

If you test this when $n = 5$, you have $_5C_4 = 5$; $_5C_2 = 10$, hence, when there are 5 points on the circumference, there are 16 regions. When $n = 6$, then $_6C_4 = 15$, $_6C_2 = 15$, and hence there are 31 regions.

Before they start step 3 in the student book, it would be a good idea to present students with some flawed proofs, and ask them to analyse the reasoning to find where the fallacy, or incorrect result, arrived at by apparently correct reasoning, occurs.

TEACHING IDEA 10: Proof that $1 = 2$

$$ab = a^2 \qquad (1)$$
$$ab - b^2 = a^2 - b^2 \qquad (2)$$
$$b(a - b) = (a + b)(a - b) \qquad (3)$$
$$b = a + b \qquad (4)$$
$$b = 2b \qquad (5)$$
$$1 = 2 \qquad (6)$$

The incorrect step is between [3] and [4] when you divide by zero! Similarly flawed reasoning can be used to show that $0 = 1$, or any number equals any other number. Try it!

Students are now ready to try step 3 in the student book.

EXTENSION 1

Suggested example for discussion:

Kati and Upol consider the sum $1 - 1 + 1 - 1 + 1 - 1 + \cdots$. Kati claims the sum is 0 because $1 - 1 + 1 - 1 + 1 - 1 + \cdots = (1 - 1) + (1 - 1) + (1 - 1) + \cdots = 0 + 0 + 0 + \cdots = 0$ but Upol thinks the sum is 1 because $1 - 1 + 1 - 1 + 1 - 1 + \cdots = 1 - (1 - 1) + (1 - 1) - (1 - 1) + \cdots = 1 - 0 + 0 - 0 + \cdots = 1$.

Who is right? What can you conclude?

TIP

Neither of them is correct as the series diverges and in fact you could group the terms in different ways to obtain whatever number you want.

EXTENSION 2

Use some famous series to write a program that approximates π to a chosen number of decimal places for example,

$$\pi = \sqrt{6\left(1 + \frac{1}{4} + \frac{1}{9} + \frac{1}{16} + \cdots\right)}$$

or $\pi = 4\left(1 - \frac{1}{3} + \frac{1}{5} - \frac{1}{7} + \cdots\right)$

Transformations and graphs

Students need to master the skill of looking at the equation of a function and knowing how it has transformed from the original root function, and hence being able to graph it efficiently. Its importance is highlighted in criterion C, in which a student's ability to move between the algebraic and graphical representation (of a function) is assessed.

With adequate scaffolding, students can discover how basic transformations occur and determine that this set of rules applies to all functions. In this activity, students have to select their own values to see how the parameters affect the root function, but you can always give these values to the students if you think it appropriate. The teaching idea below is a similar task, looking at the transformation of quadratic functions, with slightly different instructions and ideas on how to make the generalizations.

TEACHING IDEA 11: The function $y = ax^2$
Use a dynamic geometry program for this activity.

1. Open your dynamic geometry program. Use the menu GRAPH to define a new parameter a. Graph the function $y = ax^2$. ANIMATE the parameter.

   What is the effect of a on the graph? Write down your conjecture and test it further.

2. Define a new parameter c. Graph the function $y = ax^2 + c$. ANIMATE the parameter c. What is the effect on the graph of changing the value of c?

   Try to use words such as "translate", "left", "right", "down", "up", "units".

   Find the coordinates of the vertex of the graph in each case. You may need to express it in terms of c.

3. Define a new parameter b. Graph the function $y = ax^2 + bx + c$. ANIMATE the parameter b.

   What is the effect of changing the value of b on the graph?

   Try to use words such as "translate", "left", "right", "down", "up", "units".

TIP

The activity in the student book can be used as a template to conduct any transformation of a function. Simply change the root and general forms of the equation to a different function type.

TIP

Most GDCs allow the use of sliders that can be used to explore the effect of parameters by just dragging the sliders to vary their values.

Mathematicians enjoy playing with mathematical ideas just for the fun of it! Like a piece of art or music, mathematical ideas need not have any other meaning than the pleasure they give. Then, centuries later, it could happen that such playing turns out to have enormous scientific value. There is no better example of this than the work done by the ancient Greeks on the curves known as conics: the ellipse, the parabola and the hyperbola. It took almost 1800 years for mathematicians to see their practical use.

TEACHING IDEA 12: Geometrical properties of conics
TASK A: THE WORK DONE BY THE ANCIENT GREEKS

- Find who wrote about conics first.
- Find why conics have this name.

TASK B: DETAILS ABOUT PARABOLAS AND HYPERBOLAS

- Definition(s).
- What is the focus? What is the directrix? State any other elements needed to describe these conics.

TASK C: DRAWING A PARABOLA AND A HYPERBOLA

- Find at least one method of drawing these curves.

TASK D: APPLICATIONS OF PARABOLAS AND HYPERBOLAS

- Give examples.
- By referring to the properties of these conics, discuss the importance of these curves in real life.

CHAPTER LINKS

An example that you might want to highlight to students for the last bullet point in task D is the parabolic whispering wall, described on page 86 of this book, in Chapter 10 on Model. To describe why the phenomen on works, students have to discuss the focus and directrix of the parabola.

Activities 7–12 Vertical stretches, Reflections and Translations

In Topic 3 in the student book, students look at the transformation of the function $y = \dfrac{1}{x}$, a hyperbola. When this topic is combined with the investigation on parabolas on the previous page, students will have explored two conic sections that will now be the basis of a research project.

Ask students to use books or the internet to find and select information about two of the conic curves—the parabola and the hyperbola.

- The final project may take any form: hand-written, a poster presentation or a slide show, but it must provide information for each task A to D.
- Sources must be quoted and properly referenced.
- Ask students to include a brief evaluation of their work.

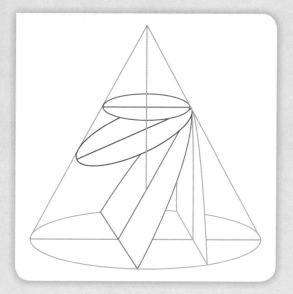

TEACHING IDEA 13: Reflecting a quadratic function

Let students complete this task.

Determine the effect of a reflection in the x-axis, the y-axis and the line $y = x$ on a quadratic function. Follow these steps.

a) Use dynamic geometry software or a GDC to create the parabola $y = (x - 4)^2$.

b) Create a table of values for this function with a domain of $-10 < x < +10$.

c) Determine the equation of each of the reflection lines.

d) Graph the three lines in which you will reflect the parabola, each on separate sets of axes.

e) Graph the original parabola on each as a point of reference.

f) Determine what the new graphs will look like reflected in each of these three lines separately.

g) Graph the three new graphs.

h) Create a table of values for each of the new graphs and compare to the original parabola.

i) Write a general statement about what happens to a parabola when it is reflected in the x-axis, the y-axis and the line $y = x$.

j) Determine the equations of the three new graphs that you created.

k) Can these generalizations be applied to other functions? Explain your reasoning. Show examples of other types of functions as validation.

TIP

Investigate reflections in different axes and the line $y = x$.
Reflecting a function in different lines or axes is like putting a mirror down on the graph and plotting all of the points of the function on the other side of the mirror exactly the same distance away.

Summary

In this chapter students explored a wide variety of problems, from number patterns, discrete graphs, geometry and graphs of functions. Using Pólya's heuristics, students were guided through several problem-solving strategies until they could reach a conclusion and produce a general statement or rule. Students were also made aware of situations that suggest a pattern that turns out to be incorrect and of the need to test further any conjectures. The activities in Chapter 8 on Justification and Chapter 11 on Pattern complement well the work suggested in this chapter. All these activities should help your students to improve their problem-solving skills and discover the joy of tackling new problems.

8 Justification

Valid reasons or evidence used to support a statement

	ATL skills	Mathematics skills
TOPIC 1 Formal justifications in mathematics		
Activity 1 Direct and indirect proofs	✓ Practise flexible thinking–develop multiple opposing, contradictory and complementary arguments.	**Algebra** ✓ Use basic algebra to conduct direct proofs. ✓ Factorize expressions to make a proof by counter-example.
Activity 2 Visual proofs	✓ Practise flexible thinking–develop multiple opposing, contradictory and complementary arguments.	**Algebra** ✓ Use linear functions and convergence of geometric series to conduct a visual proof. **Geometry and trigonometry** ✓ Use basic geometry and vertically opposite angles to conduct direct proofs. ✓ Use the properties of equilateral triangles to conduct a visual proof. ✓ Use circle and line theorems to conduct a visual proof.
TOPIC 2 Empirical justifications in mathematics		
Activity 3 Predicting the eruptions of Mount Vesuvius	✓ Interpret data.	**Algebra** ✓ Represent a piecewise-defined function both graphically and algebraically. **Statistics and probability** ✓ Use linear regression and a line of best fit to make predictions and empirical justifications.
TOPIC 3 Empirical justifications using algebraic methods		
Activity 4 Space shuttle safety	✓ Consider ethical, cultural and environmental implications.	**Algebra** ✓ Graph a rational function and use it to make an empirical justification.
OTHER RELATED CONCEPTS	Patterns Generalization Representation Change System	

Introducing justification

In Topic 1, students consider the different ways to conduct a proof. The skills of making formal justifications in mathematics and the art of conducting formal and informal proofs definitely have their place in the Middle Years Programme. Proof is not an individual separate topic; a proof is a means of communicating a logical line of reasoning and a tool used when studying mathematics in general. Students need to focus on how they will achieve this, by using clear sentences in a logical progression, detailed diagrams and/or appropriate mathematical notation and terminology.

In Topics 2 and 3 they move on to using statistical and algebraic methods of empirical justification. The ability to make generalizations and predictions in a given context and using mathematical evidence to support and/or defend a conclusion or argument are valuable life-long skills.

> *It is by logic that we prove, but by intuition that we discover.*
>
> Henri Poincaré

TOPIC 1 Formal justifications in mathematics

At the start of this section, you could tell your students the suspenseful story of Fermat's Last Theorem. It emphasizes that proving a mathematical result can sometimes be a long and weary process. In the mid-1600s, the mathematician Pierre de Fermat conjectured that the equation $x^n + y^n = z^n$, where x, y, z and n are positive integers, only has a solution for $n = 1$ or $n = 2$. In the case of $n = 1$, the equation merely states that the addition of two positive integers is a positive integer. In the case of $n = 2$, this leads to Pythagoras' theorem, for which many proofs exist. Fermat was not able to find any solutions to this equation for $n > 2$. He wrote, in the margin of the page where this conjecture can be found (in his book *Arithmetica*), that he had a proof to his conjecture, but the proof was too long to fit in the margin. For over 350 years, mathematicians attempted to prove this conjecture until finally, in 1995, after many years of working on the proof, Sir Andrew John Wiles, a British mathematician, published his famous proof of Fermat's Last Theorem.

WEB LINKS
Visit www.bbc.co.uk and search "Horizon 1995–1996 Fermat's last theorem" for an award-winning documentary on Wiles' long journey toward his proof.

TEACHING IDEA 1: Your age in chocolate
You could use this task as a preview to the chapter.

Work in a small group.

How many times a week do you eat a treat such as a chocolate bar or piece of cake? (For this task, the number must be more than one but less than ten.)

- Multiply this number by 2. Add 5 to the result.
- Multiply this number by 50. Add 1750 to the result.
- Add the last two digits from the year you last had a birthday. For example, if your last birthday was in 2014 add 14, if your last birthday was in 2013 then add 13.
- Now subtract the four-digit year that you were born. You should now have a three-digit number.

The first digit is your original number—how many times you eat a treat in a week. The second two digits are your age.

Use algebra to justify why you will always get the above result.

There are many forms of proof and some are very basic. Starting with such a basic activity can alleviate student concerns about the abstract nature of proofs.

Encourage students to write the proof in simple terms.

Let x represent the number of chocolate bars and y represent your age.

For the first few steps:

$50(2x + 5)$
$= 100x + 250$
Now, $250 + 1750 = 2000$

Adding the last two digits from the year of your last birthday (a) will always give you the current year so when you subtract the year you are born (b), you will always get your age, that is, $y = 2000 + a - b$.

So working through the proof you get:

$$50(2x + 5) + 1750 + a - b = 100x + 250 + 1750 + a - b$$
$$= 100x + 2000 + a - b$$
$$= 100x + y$$

As the number of chocolate bars is in the hundreds places, it will precede your age and you will get the three-digit answer.

Interesting points for discussion include the restrictions on age and the number of chocolate bars.

- Age cannot be over 99, as a three-digit number would carry over into the hundreds column and will affect the $100x$. The original question can be altered to accommodate people over 100 by increasing the number of place values—multiply the original number by 20 instead of 2.
- The number of chocolate bars can be greater than 10, as the hundreds column is the most significant column, so a carried figure could be accommodated. It depends if your answers can have more than three digits.
- The number of chocolate bars can be zero, but you will only end up with the two-digit age, you will not get three digits.

 Activity 1 **Direct and indirect proofs**

STEP 1 **Direct proof using geometry**

It is important that students understand the difference between inductive and deductive reasoning, and that the only method of reasoning allowed in mathematics to prove general statements is deduction. Experimentation enables them to formulate a conjecture, but deduction allows them to prove the conjecture.

The type of reasoning that involves making theories based on experimentation is called induction, and is used in the natural and social sciences. Using induction, scientists make theories that are well supported by experimental evidence, but they can never claim that these theories are "absolutely true". Indeed, such theories are often found to be faulty, and are replaced by new theories when there is new experimental evidence.

Dynamic geometry software is an excellent tool for learning inductive reasoning. It allows students to experiment and formulate a conjecture in an efficient manner. To prove their conjectures, students need to be given a list of axioms to work from. This list will grow as they prove more of their conjectures.

The process described here is a good introduction to direct proofs, using geometry, and can be used as a building block to create a list of axioms to use for further proofs. Start by giving students two theorems to work with and, through deductive reasoning they prove more theorems, which they can use in further proofs.

a) Formulate a conjecture about the angle sum of a triangle. Students will already know this but you have to tell them it is only a conjecture until it is proven.

b) Use the theorems that alternate angles are equal and the sum of adjacent angles forming a straight line is equal to $180°$ to prove the conjecture that the sum of the angles of a triangle is $180°$.

Note: When students have completed the proof in **b)**, they can tackle step 1 in the student book.

c) Formulate a conjecture about the relationship between the exterior angle of a triangle and the sum of the interior opposite angles.

d) Use proven theorems to prove that the exterior angle of a triangle is equal to the sum of the interior opposite angles. Students can add the angle sum of a triangle proof to their list of theorems and then use it in this example.

Students can then use DGS to formulate conjectures and try to prove them, using their ever growing list of theorems.

STEP 2 **A direct proof using algebra**

Algebraic derivations and proofs are useful and accessible to students of this age, and will be very helpful in the Diploma Programme. In particular, after they have completed the simpler ones in the student book, you could show them the derivation of the quadratic formula. Other useful derivations and proofs, particularly as preparation for the DP, are sums and products of roots of quadratic equations, and the factor and remainder theorems for polynomials. Often, the first obstacle is to translate the language words into mathematics but, with a bit of practice using the examples in the student book, this too will become easier.

A few interesting algebraic proofs are suggested here.

a) Prove that every odd integer is the difference of two perfect squares.

b) If a is an integer divisible by 4, then a is the difference of two perfect squares.

c) Prove that for any real numbers a and b, $a^2 + b^2 \geqslant 2ab$.

STEP 3 **Indirect proof (proof by contradiction)**

Students will need guidance in setting up the proof by contradiction, particularly in stating which part of the statement will be assumed. Below are two classic proofs for additional work on using this method of proof.

a) There are no positive integer solutions to the equation $x^2 - y^2 = 1$.

By inspection, students can easily arrive at $x = \pm 1$ and $y = 0$ as a solution to the equation. Alternatively, students assume that there is such a solution, hence, $(x+y)(x-y) = +1$. This means that $x+y = 1$ and $x-y = 1$, or $x+y = -1$ and $x-y = -1$. Solving simultaneously they arrive at the same answer.

b) Students can now use the method of indirect proof, together with the result in step 3 of the activity in the student book, to show that $\sqrt{2}$ is an irrational number. That is, $\sqrt{2}$ cannot be expressed in the form $\frac{p}{q}$, where p and q are integers, $q \neq 0$, where the fraction $\frac{p}{q}$ is completely simplified.

TIP

Another way of describing mathematically that the fraction $\frac{p}{q}$ is completely simplified is to state that the gcd $(p, q) = 1$. This reads as the greatest common divisor of p and q is one, that is, p and q are relatively prime.

c) They can use the method of proof by contradiction to prove that there are an infinite number of prime numbers.

d) They could research how the result in **c)** is used in real life, for example, in cryptography and in creating bar codes.

STEP 4 Proof by counter-example, or disproof

Students sometimes need convincing that this is a valid method to disprove a statement. It is perhaps easiest to start with an example within the classroom for example, "All students in this class have red hair."

For the group activity in step 4 of the student book, the first value of n for which the coefficients of the variables of the factors of $x^n - 1$ are not ± 1 is $n = 105$. This is an excellent example for understanding that, in mathematics, you can never stop after what you consider to be a reasonable number of individual examples that confirm a conjecture. Rather, it is essential to prove the conjecture deductively, using either a direct or indirect method, or to find a counter-example to disprove the conjecture.

TEACHING IDEA 2: Proving statements

Ask students how they would prove the statement that all students in this class are brunette. Isn't it sufficient to find one student that isn't brunette in order to disprove the statement? Ask for more examples from their everyday lives, and how they use this method many times, even if not consciously.

 Activity 2 **Visual proofs**

Visual proofs provide an opportunity to develop good mathematical communication. It is important to look at all the details on the diagrams and discuss their relevance. In some cases it is necessary to make assumptions explicit.

The first visual proofs task in the student book includes a detailed diagram without any hidden assumptions.

STEP 1 The island problem

Mrs Robinson's drawings show three equilateral triangles and their heights. To justify that the hut can be built anywhere on the island, students need to notice these points.

- The heights with respect to any vertex in an equilateral (regular) triangle are equal; moreover if all the altitudes are drawn, the segments meet at the centre of the triangle.
- If the two top triangles are rotated about their centres the distances from each point to the sides of the triangle can be viewed in the vertical position.

- The small triangle shown on the diagram can be translated along the side of the big triangle. This transformation preserves the lengths.
- The second diagram shows clearly that the sum of the three distances to the sides of the big triangle is equal to its height.

So, no matter where the hut is built, the sum of the distances to the sides of the triangle is constant and equal to its height.

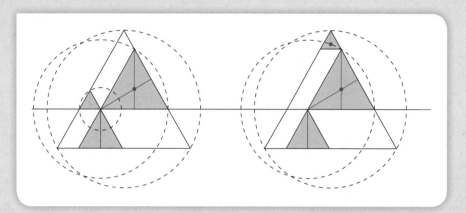

STEP 2 **Adding to infinity**

The second step is more demanding. It is important to discuss the condition under which two lines meet in the first quadrant. As one of the lines is fixed, the conditions will need to be imposed on the other line.

The diagram shows two straight lines with equations $y = x$ and $y = rx + a$, $r > 0$. Assuming that these lines meet at a point (x, y) with $x, y > 0$, this implies that the gradient (slope) of $y = rx + a$ needs to be smaller than the gradient of $y = x$, that is, $r < 1$.

Solve the equations simultaneously.

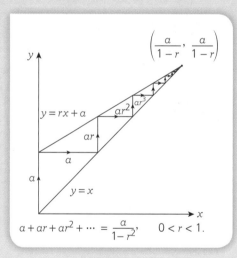

$$\begin{cases} y = x \\ y = rx + a \end{cases} \Rightarrow \begin{cases} y = x \\ x = rx + a \end{cases} \Rightarrow \begin{cases} y = x \\ (1 - r)x = a \end{cases} \Rightarrow \begin{cases} y = x \\ x = \dfrac{a}{1 - r} \end{cases}$$

Therefore, if $0 < r < 1$, the lines meet at $P\left(\dfrac{a}{1 - r}, \dfrac{a}{1 - r}\right)$. Now, starting from the origin, you can move horizontally and vertically to point P.

From the origin, move along the y-axis to the point $(0, a)$ where the line $y = rx + a$ meets the y-axis; then move horizontally until you meet the line $y = x$; then vertically until you meet the line $y = rx + a$, and so on as illustrated in the diagram. Now you just need to note that the x-coordinate of P must be equal to the sum of all horizontal segments: $a, ar, ar^2, \dots$ (and the y-coordinate must be equal to the sum of the vertical segments) between the lines.

To obtain the lengths of these segments you just need to compare the x-values of the points on the lines for the same value of y (and compare the y-values of the points on the lines for the same value of x).

The infinite series has a finite sum because $0 < r < 1$.

CHAPTER LINKS
Geometric series are explored in more detail in Chapter 5 on Change.

The last task on visual proofs requires knowledge of the circle theorem: an angle inscribed in a semi-circle is a right angle. This result needs to be introduced or revised before students attempt this proof by contradiction.

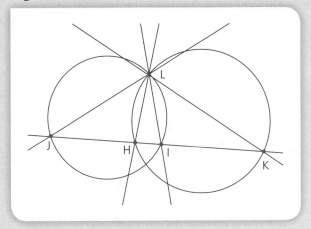

Suppose that H and I are distinct points. Then both $\angle JIL$ and $\angle KHL$ are right angles as these are angles inscribed in a semi-circle, since JL and LK are diameters. But then $\angle LHI + \angle HIL = 90° + 90° = 180°$. But this means that the triangle LHI does not exist as $\angle HLI = 0°$. Therefore H and I must be the same point. This means that if two circles intersect at a point L and LJ and LK are the diameters of these circles through L, then the line JK meets the circle at their other intersection point as shown below.

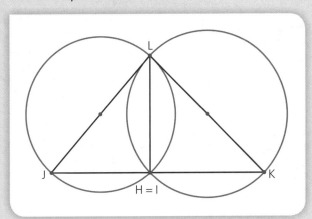

TOPIC 2   Empirical justifications in mathematics

In this topic students use mathematical methods to draw conclusions and make decisions that they can justify.

👤 Activity 3   Predicting the eruptions of Mount Vesuvius

The table in the student book gives data from the Pompeii eruption up to the early twentieth century. Students will interpolate and extrapolate data to find information about other eruptions. The task is suitable as a summative task for a unit on linear functions.

Assessment

If you choose to assess students on this task, you can use criterion D.

The task-specific descriptor in the top band (7–8) should read that the student is able to:

- select an appropriate method to determine the equation(s) of the model(s)
- determine the equation(s) of the model(s) correctly
- determine the domain of model(s) correctly, where necessary
- use the model(s) correctly to predict the year of the fifth and last eruptions
- justify the degree of accuracy of the predictions
- justify whether the predicted values make sense in the context of the problem.

TIP

The data for Mount Vesuvius actually follow two linear patterns. Students are asked to find the year of an eruption within each pattern. They should find two lines of best fit and, therefore, two equations. They should also be indicating the domain for each equation.

Stage 1 of the unit planner

Key concept	Related concepts	Global context
Relationships	Justification Change Pattern	Orientation in space and time
Statement of inquiry		

Representing patterns as relationships can help us understand how natural landscapes change over time.

⚭ INTERDISCIPLINARY LINKS
There are many possible links to individuals and societies for this task: looking at how archeologists have used Pompeii to discover many elements of ancient Roman civilization and how civilizations evolve, or conducting an analysis of past and present volcanic activity and how societies cope with natural disasters in general.

TEACHING IDEA 3: Piece-wise defined functions

INQUIRY   Rather than simply learning about piecewise-defined functions, students could perform this activity as an introduction to the topic. Even students who have no experience with functions defined in this way will see that one linear function will not accurately describe the data. By asking students a few leading questions, such as: "What x-values would you use in each equation?", you can lead them to discover both the need and the mathematics of piecewise-defined functions.

Students can choose their preferred type of regression analysis to find the line of best fit, including placing it by eye. Finding the equation poses few problems. Identifying the domain is more difficult. Ask simple questions such as: "For what x-values would you use each equation?" and "How would you describe that mathematically?" Make sure that students do not assign domains that overlap and that they use appropriate mathematical notation.

EXTENSION

After students have tackled a linear regression based on a piecewise function, they could try working with another data set that needs to be broken down into sections in order to make predictions.

The following task looks at the strengths and limitations of regression models.

Instructions to students

Answer all the questions. Use algebra and your graphic calculator. Input calculator images where needed.

STEP 1   The concentration (in micrograms per millilitre) of pain-relieving medication in a patient's blood was monitored every ten minutes for two hours. The table gives the data.

TIP

Drug concentration in the bloodstream typically follows quite a complex rational function but as it is difficult to model data with rational expressions, the students will find it easier to break the data into sections and treat it as separate quadratic and exponential functions.

Time (minutes)	0	10	20	30	40	50	60	70	80	90	100	110	120
Concentration (μg/ml)	0	1	4	7	12	16	18	19	22	24	25	26	25

Use a GDC to create a scatterplot of the data.

Determine the equation of the line of best fit. Justify your choice, showing all your statistical and algebraic reasoning.

Does the shape of the graph of your equation of best fit make sense? Explain, given the context of the question.

Using your equation, determine the concentration at 15 minutes.

Using your equation, at what time was the concentration at 14 μg/ml?

On the medication's label, the directions read that it can be taken every 4–6 hours. Does that sound reasonable? Explain why or why not.

Do you feel your equation of best fit is accurate? Justify your answer.

Do you think that two hours was a long enough time to monitor the patient, to make predictions about the effectiveness of this medication? What would you have done?

Does the fact that you can mathematically justify a model to fit a data set make it a valid choice? Discuss.

STEP 2 This table shows data for the same patient, with the same drug concentration. This time, the results were recorded for 12 hours.

Time (hours)	0	1	2	3	4	5	6	9	12
Concentration (μg/ml)	0	18	25	21	17	13	7	3	2

How does this compare to your first graph and regression curve?
Does your new regression curve still accurately represent the data?

Try to find a function type that better fits the data. Justify your choice, showing all statistical and algebraic reasoning. (You may divide the curve into parts to create a piecewise function, making sure that the end points in the domain of each section do not overlap.)

Based on the new model, can this drug be taken every 4–6 hours? What other information would you like to know that would affect your decision?

What conclusions can you make about building regression models? Comment on their strengths and limitations.

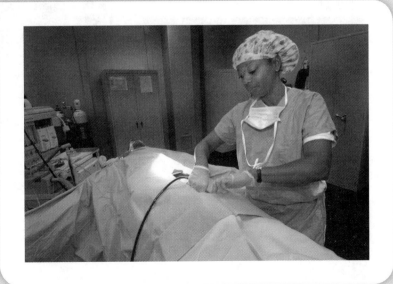

Empirical justifications using algebraic methods

In this chapter, students have had many opportunities to justify either their mathematics or their mathematical conclusions. However, sometimes conclusions and decisions are made based on mathematics. In this topic, students will use mathematics to investigate competing priorities as they justify a proposed budget for a mission into space.

Activity 4 — Space shuttle safety

This activity is appropriate as a summative task for a unit on rational functions.

Assessment

If you choose to assess students on this task, you can use criterion D.

The task-specific descriptor in the top band (7–8) should read that the student is able to:

- graph the function correctly, clearly indicating the required five points and the asymptote
- interpret the significance of the asymptote correctly, in the context of the problem
- justify the degree of accuracy of the proposed budget
- justify why the suggested budget is appropriate, given the context of the problem.

Stage 1 of the unit planner

Key concept	Related concepts	Global context
Relationships	Justification Pattern Model	Identities and relationships
Statement of inquiry		
Modelling relationships can help us to justify our values and choices.		

This activity assumes some knowledge of rational functions and their asymptotes. Students should start by graphing the function $y = \dfrac{1}{x}, x \neq 0$ so that they can be introduced to asymptotes before moving on to discovering how to identify asymptotes of rational functions of the form $y = \dfrac{ax+b}{cx+d}, x \neq -\dfrac{d}{c}$.

CHAPTER LINKS

There is an activity on the transformations of basic rational functions in the "Generalization" Chapter 7 of the student book that could be used as an introduction to rational functions.

TIP

While vertical asymptotes typically make sense to students, they often wonder what a horizontal asymptote is. While students often hear that "it's an invisible line that a graph does not touch", this isn't entirely true. Students need to look at the "end behaviour" of a graph or, potentially, the notion of a "limit" to understand the real meaning of a horizontal asymptote. At this point, students are ready to learn how to graph a rational function, first starting with its asymptotes.

TEACHING IDEA 4: Drawing a graph

INQUIRY Let students use a graphic display calculator (GDC) to draw the graphs of these rational functions and copy each one into their notebook. Then ask them to find the vertical and horizontal asymptotes for each curve, from the GDC, and write their equations next to the graph. There should be no decimals or fractions (sometimes the GDC isn't clear).

a) $y = \dfrac{x+3}{x-2}$ b) $y = \dfrac{4x-1}{x+3}$ c) $y = \dfrac{6x-12}{2x-2}$

Ask students to identify a pattern for each of the equations of the horizontal and vertical asymptotes. How do they relate to the given function?

Encourage students to guess the asymptotes for these functions before graphing them. They should still draw each one on their GDC, copy the graph and identify the equations of the asymptotes.

d) $y = \dfrac{2x+8}{2x+4}$ e) $y = \dfrac{8x}{4x-12}$ f) $y = \dfrac{x+3}{x-2}$

Based on their results, ask them to think of an easy way to determine the asymptotes without actually graphing the function.

In part **c)** of this activity, students should quickly realize that the horizontal asymptote is 100, meaning there is no guarantee of success. There will always be some chance of failure. They will then need to decide what level of uncertainty is acceptable, to determine the budget for safety. What started out as a mathematics problem has now also become one of ethics and values: What is an acceptable level of security? What if reaching that level costs too much? What price can you put on safety?

Summary

This chapter started with students looking at formal justifications. They used axioms and theorems to create a sequence of logical arguments that prove if their mathematical conjecture is correct. The tasks ranged from using simple algebra or geometry in direct proofs, to indirect proofs involving contradiction and counter-example. Students then used visual models to conduct visual proofs and converted them into formal proofs, in which they justified all the steps, using results that are previously proved to be true.

The students then moved on to empirical justifications. For these they used algebraic and statistical methods and linear regression and piecewise functions to make predictions about real-world topics (in this case, eruptions of Mount Vesuvius) and justify their calculations. Students also graphed a rational function to make informed decisions about real-world topics (in this case the balance between monetary budget and astronaut safety) and justify their position.

While these approaches to justification are all very different, each is equally important in developing students' ability to make generalizations (and predictions) and justify their position or conclusion, which are valuable lifelong skills.

CHAPTER

9

Measurement

A method of determining quantity, capacity or dimension using a defined unit

	ATL skills	Mathematics skills
TOPIC 1 Making measurements		
Activity 1 Investigating rulers	✓ Apply existing knowledge to generate new ideas, products or processes.	**Number** ✓ Solve problems in context, using ratios and proportions. ✓ Make conjectures based on measurements.
Activity 2 Up, up and away!	✓ Consider multiple alternatives, including those that might be unlikely or impossible.	**Number** ✓ Solve problems in context, using ratios and proportions. **Geometry and trigonometry** ✓ Solve problems in context using formulas for volumes of 3D geometric shapes.
TOPIC 2 Related measurements		
Activity 3 Tangents to circles	✓ Test generalizations and conclusions.	**Algebra** ✓ Solve rational equations. **Geometry and trigonometry** ✓ Discover the relationships between central and inscribed angles in a circle. ✓ Discover the relationship between the lengths of tangents to a circle. ✓ Determine lengths and lengths of objects, using similar triangles. ✓ Conduct geometric proofs.
Activity 4 Lengths of segments of chords	✓ Test generalizations and conclusions.	**Geometry and trigonometry** ✓ Discover the relationships between segments in a circle. ✓ Conduct geometric proofs.
Activity 5 Points external to the circle	✓ Test generalizations and conclusions.	**Geometry and trigonometry** ✓ Discover the relationships between segments in a circle and tangents to a circle. ✓ Conduct geometric proofs.
Activity 6 Cyclic quadrilaterals	✓ Test generalizations and conclusions.	**Geometry and trigonometry** ✓ Explore the properties of cyclic quadrilaterals. ✓ Apply the relationships between central and inscribed angles in a circle. ✓ Conduct geometric proofs.

Activity 7 The circumference of the Earth	✓ Apply skills and knowledge in unfamiliar situations.	**Number** ✓ Solve problems in context, using ratios and proportions. **Algebra** ✓ Solve rational equations. **Geometry and trigonometry** ✓ Determine lengths and heights of objects, using similar triangles and trigonometric ratios.
Activity 8 Measuring the immeasurable	✓ Organize and depict information logically.	**Geometry and trigonometry** ✓ Determine lengths and heights of objects, using similar triangles and trigonometric ratios. ✓ Solve problems in context using sine rule and cosine rule.
OTHER RELATED CONCEPTS	**Generalization** **Representation** **Justification**	

Introducing measurement

Measurement is one of the most basic concepts in mathematics and is fundamental to many areas of both mathematics and science. From measurements, conjectures can be made that can later be proven and established as theorems. These can be used to find new measures, even of objects that cannot easily be measured with the tools that are readily available. In this chapter, students will not only look at the tools of measurement but also consider how they can be used to create and extend their knowledge of mathematics and the world around them.

> *Nearly all the grandest discoveries of science have been but the rewards of accurate measurement and patient long-continued labour in the minute sifting of numerical results.*
>
> Baron William Thomson Kelvin

TOPIC 1   Making measurements

When approaching this topic, consider working with other measurements, such as temperature. The teaching idea below is an experiment that students can perform to analyse the insulation capabilities of typical cups used to serve hot beverages: What is the best insulator for a hot drink—paper, styrofoam or ceramic?

Teaching idea 1 describes an experiment in which students boil water and take measurements as it cools. If this is not a viable activity for you to do in your classroom, due to lack of access to a laboratory or health and safety concerns, you could gather the data by doing the experiment as a demonstration in front of the students. Alternatively, you could simply give them a table of the data to work with. Keep in mind that the focus of this activity then moves towards mathematical modelling and away from measurement.

∞ INTERNATIONAL MATHEMATICS

Different countries may serve hot food or drinks in containers made from a range of materials. If it is appropriate, substitute and test whatever material you choose, such as glass or hard plastic.

It will take a significant amount of time (at least 30 minutes and up to 1 hour) to get enough data points for students to determine the exponential regression model to represent the cooling of the hot liquid. Make sure you have other activities for the class to do.

Students should have studied exponential functions by this stage so that they will recognize the shape when they draw the graph. They can compare it to other functions they have studied, including linear, quadratic and perhaps logarithmic and rational functions.

CHAPTER LINKS

The equation for the linear relationship between Fahrenheit and Celsius is covered in the student book, in Chapter 13 on Representation.

INTERDISCIPLINARY LINKS

You could also do this as a possible interdisciplinary activity with science. The experiment could be done more formally in the laboratory, using the experimental method. The students could bring the raw data to their mathematics class, where you could complete the analysis with them.

TEACHING IDEA 1: Testing insulators

Materials

In order to conduct this measurement experiment, students will need:

- insulators to test—paper, styrofoam, glass and/or ceramic mugs (some questions below consider take-away containers so paper and styrofoam must be included in the list)
- a hot plate or a kettle to boil the water
- a thermometer for each container
- a stopwatch or other device to record time (Have a discussion with the students about the accuracy needed for this experiment.)
- a copy of the method and instructions.

Method

Measure out exactly the same amount of boiling liquid for each container. Place a thermometer in each container and record the temperature at time zero. Every 2–5 minutes, record the temperature of the liquid in each container.

When the measurements are complete, graph the data, on paper or by using a GDC, and then determine a mathematical model that would best represent the relationship between time and temperature. Determine a regression model equation to represent the data and justify your choice of function.

Extension

Follow up with questions such as these.

- Will the temperature of the cooling liquid in the classroom ever reach 0°C or 32°F? Explain.

- The types of function being investigated generally have horizontal asymptotes. What would the asymptote be in this experiment? Why?

- Would it matter if your units were in Fahrenheit or Celsius? Explain.

  a) Based on the experiment, which is the best insulator for a cup of coffee (or hot liquid)? Justify your choice.

  b) Do cafes and restaurants use this material for their take-away coffee mugs? If not, give reasons why they choose a different material.

  c) What other factors would affect the cooling rate of the hot liquid? Are these taken into consideration when you look at the design of the take-away coffee mug?

MATHS THROUGH HISTORY

There is a famous lawsuit that occurred in the 1990s, regarding a scalding hot cup of coffee, that caused much debate. In 1992, a woman in New Mexico, USA, bought a cup of coffee at McDonald's and spilled it on her lap. McDonald's had a policy at the time to serve coffee at 180–190°F (82–88°C), which could cause third degree burns in under 10 seconds. She sued McDonald's and a jury awarded her nearly $3 million in punitive damages for the burns she suffered. The woman's attorney argued that coffee should never be served hotter than 140°F (60°C) to avoid such horrible burns. This is the average temperature of a typical home-coffee brewing machine.

Activity 1 Investigating rulers

In this activity, students explore special kinds of ruler that have minimal markings on them. For example, given a ruler of length 7 units, is it possible to measure every length up to 7 without having the ruler divided into seven units? This is a useful problem-solving activity that is accessible to all students, since very little prerequisite knowledge is required. Encourage students to take a systematic approach and tabulate their results.

EXTENSION Golomb rulers

Golomb rulers are rulers with minimal markings and with which each measurement can only be made in one way. For example:

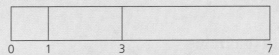

This is a Golomb ruler since it can measure the lengths of 1, 2, 3, 4, 6 and 7 units, but each can be measured in only one way. It would be a perfect Golomb ruler if it could measure all of the lengths up to 7, but this ruler cannot measure a length of 5. Students could research the work of Solomon Golomb or other Golomb rulers . They could also decide if the ones they design in the activity are Golomb rulers.

Whilst in the first activity students investigate measuring tools, this next activity invites them to take measurements and then use them to find other measures.

 Activity 2 Up, up and away!

The goal of this activity is, in the absence of any guide or scale, to find an appropriate yardstick, or object to use, to make a comparison. The student can then use this to scale the object, to find certain measurements. An important part of the process is a discussion about the accuracy of their yardstick and their final results.

Students will probably want to use something such as the height of a building or a person as a yardstick. They could research the average height of an Austrian male, female or child, if they choose to use a person.

EXTENSION The giant Easter egg

Estimating areas and volumes of unfamiliar shapes is an excellent task to improve problem-solving skills. When they are done as group activities, they provide the opportunity for students to improve their mathematical communication. For such activities, newspapers, magazines or photos can provide ample shapes to do geometric modelling. For example, students may be asked to find the volume of the giant chocolate Easter egg that took 26 Belgian chocolatiers to make! (Search online for "world's largest Easter egg by Guylian".) They should first find an appropriate yardstick, such as a person in the Easter egg photograph, and then decide on a scale to find the dimensions of the egg. They may use one geometric shape, or consider splitting the egg into different geometric shapes in order to improve the accuracy of their answer.

> **TIP**
>
> It would be advisable to give students a list of volume formulas for different 3D shapes, for example, cone, frustum of a cone and sphere.

REFLECTION

a) What is the effect on your calculations of modelling the balloon as a sphere?

b) Describe at least two other ways to improve the accuracy of your result. Be specific.

c) What difference does an error of 10 cm in the height of the person make to the overall balloon volume?

Related measurements

This topic leads students to establish relationships among segments and among angles in circles. Based on measurements, students arrive at conjectures and then are asked to use previously learned theorems to prove what they have found. This is a good way of mixing the practical with the theoretical, as students discover that they can find rules and then prove them. The rule is usually quite easy for them to find, especially when using dynamic geometry software for their measurements. Where they will struggle, however, is in the proof of their conjectures.

Activity 3   Tangents to circles

Students will discover that the lengths of tangents to a circle from a point exterior to the circle are equal. This means that, during an eclipse, the distances to the top and bottom of the Moon and/or Sun are equal only from the cities that experience the full eclipse.

The proof of this theorem is actually quite simple, once students draw the radius from the centre to the point of contact of the tangent. Then they can use the result from **c)** in the teaching idea below, with Pythagoras' theorem.

TEACHING IDEA 2: Circle theorems

INQUIRY Most of the theorems in circle geometry can be developed through inquiry-based learning activities, such as those in the student book. Here are some further suggestions for involving students in the discovery of mathematics. Encourage students to complete these before attempting those in the student book since some of their proofs rely on these results. Through these activities your students will arrive at conjectures only. They should either be shown or asked to develop a geometric proof for each result.

a) Instruct your students to draw circles and chords and construct perpendicular bisectors to verify that the perpendicular bisector of a chord passes through the centre of a circle. Next, ask them to use the same method to show that the bisector of a chord that passes through the centre of the circle must be perpendicular to the chord. Investigate the other variations on that rule.

b) Instruct the students to draw circles and then construct angles at the centre and at the circumference (central and inscribed angles), based on the same chord or arc. Now ask them to measure each angle to determine the relationship between angles at the centre and at the circumference.

c) Now ask the students to draw more circles and add tangents to them at various places. Ask them to draw a radius from the centre to each point of contact, where the tangent touches the circle (the point of tangency). Now ask them to measure the angle formed by the tangent and the radius.

d) Ask students to draw a tangent to touch a circle where a chord meets the circumference. Ask them to construct an angle at the circumference, based on the endpoints of the chord (the inscribed angle that contains the chord). They can now measure each angle and compare them. (This is know as the "angles in the alternate segment theorem".)

TIP

It is important that once students see the relationship, they formulate it in their own words. They can then be shown how others have formulated the same result and subsequently be asked to construct a formal proof. To help them remember this new information, give students opportunities to practise and apply what they have learned. Thus, they go from constructing knowledge to using it and then back to constructing and using more knowledge, and so on.

∞ WEB LINKS

Proofs of the theorems in this section can be found at http: //m. everythingmaths.co.za. Go to **Read Maths 12 > Analytical Geometry.**

TEACHING IDEA 3: Sentence stems and incomplete sentences

Each of these activities requires students to formulate their conjecture in a written form before proving it geometrically. If students struggle (with either language or mathematics), give them sentence stems or incomplete sentences that they simply need to complete. The following examples could be used for the four activities in this topic.

Activity 1:　The lengths of tangents to a circle from a point _____ .

Activity 2:　The _____ of segments of one chord is _____ to the _____ of segments of another _____ that it _____ .

Activity 3:　If two secants share an endpoint not on the circle, _____ .

　　　　　　If a secant and a tangent share an endpoint _____ .

Activity 4:　Opposite angles of cyclic quadrilaterals _____ .

Solution

Triangle ESA and triangle EMB are similar because all three pairs of corresponding angles are the same (AAA). (Students should go through the formal process of showing the proofs for this.)

Students can either use geometry or trigonometry to show that:

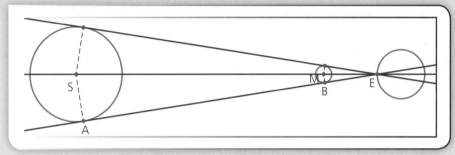

$$\frac{SA}{SE} = \frac{MB}{ME}$$

$$\frac{r_{Sun}}{d_{EtoS} + r_{Sun}} = \frac{r_{Moon}}{ME}$$

$$\frac{695\,500}{149\,600\,000 + 695\,500} = \frac{1740}{ME}$$

$$ME = 376\,000\,km$$

Therefore the furthest distance the moon can be from the Earth is $376\,000 - 1740 = 374\,260$ km.

The Moon, on average, is about 384 400 km from the Earth, with a range from about 363 100 km at its closest to 406 700 km at its furthest.

The distance of 374 260 km is not much greater than 363 100 km, the closest position the Moon gets to Earth in its orbit, and is smaller than the Moon's average orbit. So it is no surprise that eclipses are rare occurrences. Other factors also come into play, such as alignment, further adding to the rarity of a total solar eclipse.

👤 Activity 4　Lengths of segments of chords

In this activity, students should arrive at the fact that, in the circle:

$$MC \times MD = MA \times MB$$

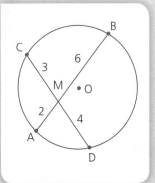

🔗 WEB LINKS

Students can access the NCTM Illuminations website and use their "Power of a Point" applet instead of drawing and measuring. There is an option to show the product of segments so that students can see what happens as point M, in the diagram, is moved.

TIP

For students who struggle, it may be necessary to use dynamic geometry software or help them measure, since testing the result relies on accurate measurements. Once again, it is appropriate to have students practise using this new information before going on to other theorems.

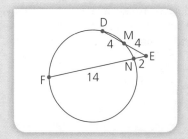

Rather than just using the diagrams in the student book, students could draw their own diagrams. Simply let them construct a circle, choose a point outside the circle and then draw two lines that each intersect the circle in two places. In that way, they don't think that the problem was specifically designed to make the relationship work. In either case, students' initial attempts are likely to involve testing if EM × MD = EN × NF. Allow them the freedom to make that mistake and learn from it, though they are likely to try it several times!

Students are often puzzled when one of the lines is a tangent instead of a chord. To prepare them for this development, ask them to construct their own problems (as described previously) but be sure to draw one line so that its points of intersection are successively closer and closer together. This allows them to see that the two measurements are now becoming roughly the same and usually helps a majority of students recognize that the distance to the point of contact of the tangent needs to be squared.

TIP

Make the connection between this theorem and the previous one. Ask students what segments were multiplied when the point was inside the circle. Make sure they notice that the distance is always from the point of intersection to the point on the circle, not from a point on the circle to another point on the circle.

Activity 6 Cyclic quadrilaterals

In this last inquiry activity, students need a protractor. The result is usually quite easy for them to see, although the proof is not as obvious. It can be helpful to let them attempt the proof, even with teacher support, so that they see how theorems can build upon one another. This proof uses the fact that central angles are twice as large as inscribed angles.

TEACHING IDEA 4: Application of theorems
Once students have learned and practised the theorems in this topic, give them the opportunity to see how they relate to real-world problems.

Student instructions

a) When hiking or skiing in areas that are not patrolled, one should always wear a transceiver or rescue beacon. This is a device that emits a signal that other transceivers can pick up as far as 30 metres away, in case of an accident or avalanche. Transceivers send out signals that travel outwards, in concentric circles, with the transceiver as the centre. How could someone with another transceiver use circle geometry to narrow down the position of the person emitting the signal?

Answer
The person with the receiver walks until they pick up the signal. They record that position, then keep on walking until the signal is lost (again recording the position). The path they have just taken, between the two points, forms the chord AB (see the diagram on the right). If the person then goes to the midpoint of the two positions measured, and walks at 90 degrees to the previous line, the emitting transceiver is along that line.

TIP

Providing this picture to students as a hint will help those who struggle.

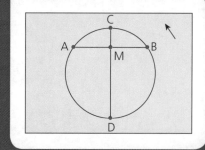

For a more specific location, the person should start at that midpoint and walk away from the signal, until they lose it. That locates point C. They then walk from C, through the midpoint (M) until they reach the other end of the circle (point D). This gives the diameter, since the perpendicular bisector goes through the centre. The location of the lost hiker (the emitting transceiver) is midway between C and D, the centre.

b) How many satellites are needed to observe all of Earth?

Every night, there are thousands of satellites circling the Earth. You may have looked up and thought you were seeing a shooting star, except that it didn't fade away! These satellites perform a variety of functions, from relaying mobile phone calls, to helping with Global Positioning Systems (GPS) to monitoring weather. Each satellite can only see a portion of the Earth at a time, so how many would you need to observe the whole Earth at one instant?

In terms of visible area, what happens as the satellite moves further and further away from the Earth?

If the Earth has a radius of 6371 km and the satellite is 35 000 km above the planet, how much of the Earth can it see?

What is the maximum angle that a satellite can cover on the Earth?

Based on your calculations, what is the minimum number of satellites needed to observe the entire Earth in one instant?

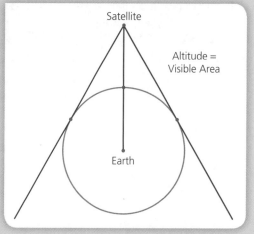

TOPIC 3 Using measurements to determine inaccessible measures

For this topic, students use geometry and trigonometry to calculate the circumference of the Earth, based on measurements made on a much smaller scale.

TEACHING IDEA 5: Basic trigonometry, the three trigonometric ratios

INQUIRY Rather than teaching students explicitly the trigonometric ratios, let them discover them by drawing triangles, measuring sides and finding ratios. This activity is very effective in accomplishing just that. Give students copies of the instructions.

Instructions

a) Using a protractor and a ruler, draw two right-angled triangles of different sizes, but with a second angle measuring 40°. On each triangle mark clearly which angle is the right angle and which one measures 40°.

b) (i) Which is the longest side of the triangle? How do you know? Write beside it the word "hypotenuse".

(ii) Write on the side across from the 40° angle the word "opposite". Why does that make sense?

(iii) On the third side write the word "adjacent". What does "adjacent" mean? Why is this an appropriate name for this side?

TIP

Once students have discovered the basic trigonometric ratios, give them ample practice before they move on to the summative task. This practice should include solving right-angle triangles, including some that require multiple steps.

c) Now work with another student and fill in the table for the four triangles you have created. You will need a ruler and a calculator for this part of the activity.

	Length of opposite side (cm)	Length of adjacent side (cm)	Length of hypotenuse (cm)	$\dfrac{\text{opposite}}{\text{hypotenuse}}$	$\dfrac{\text{adjacent}}{\text{hypotenuse}}$	$\dfrac{\text{opposite}}{\text{adjacent}}$
Triangle 1						
Triangle 2						
Triangle 3						
Triangle 4						

d) **(i)** What do you notice about the four answers to the ratio $\dfrac{\text{opposite}}{\text{hypotenuse}}$?

(ii) What do you notice about the four answers to the ratio $\dfrac{\text{adjacent}}{\text{hypotenuse}}$?

(iii) What do you notice about the four answers to the ratio $\dfrac{\text{opposite}}{\text{adjacent}}$?

(iv) Why do you think that happened? Give at least one mathematical reason.

e) Now repeat the same activity, this time drawing right-angled triangles with a different second angle (say, 65°).

f) Each of these ratios has its own name and abbreviation. Do some research and fill in the mising words.

$$\underline{\hspace{2cm}} = \dfrac{\text{opposite}}{\text{hypotenuse}}; \qquad \underline{\hspace{2cm}} = \dfrac{\text{adjacent}}{\text{hypotenuse}}; \qquad \underline{\hspace{2cm}} = \dfrac{\text{opposite}}{\text{adjacent}}$$

> **TIP**
> This can also be done with dynamic geometry software.

👤 Activity 7   The circumference of the Earth

In this activity, students recreate some of the work done by both Eratosthenes and Biruni in trying to determine the circumference of the Earth.

This task could be used as a summative assessment task.

Assessment

If you choose to assess students on this summative task, you can use criterion D. To ensure that it is appropriate for criterion D, you will need to remove some of the steps, so that students have the opportunity to "select and apply" their own problem-solving techniques. Another option is to remove that strand of the criterion. Just make sure that you assess it at least twice during the school year with other assessment tasks.

The task-specific descriptor in the top band (7–8) should read that the student is able to:

- select and apply an appropriate method to determine the radius of Earth
- correctly determine the radius of the Earth
- correctly justify the steps of their solution
- justify the degree of accuracy of the radius of the Earth
- justify whether the calculated radius makes sense in the context of the problem.

Key concept	Related concepts	Global context
Relationships	Measurement Generalization	Orientation in space and time
Statement of inquiry		
Humans have generalized and applied relationships between measurements in an attempt to define the planet on which we live.		

a) Eratosthenes' method

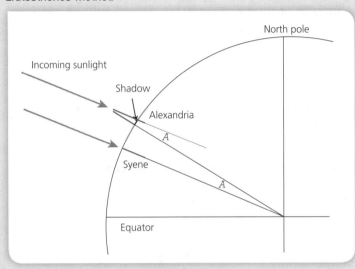

Based on the diagram above, the angles marked A have to be congruent because they are alternate interior angles. Students should have used trigonometry to calculate the angle of elevation of the Sun to be 82.8°, allowing them to find the value of A. Knowing that A is 7.2°, students should be able to set up this ratio:

$$\frac{7.2°}{360°} = \frac{5040 \text{ stadia}}{C}$$

Where C is the circumference of the Earth.

b) Biruni's method

Biruni's method uses trigonometry instead. Students may have difficulty setting up the equation, but once they do they should be able to solve it. You could ask them to solve questions such as these as a warm-up at the beginning of class.

a) $\dfrac{2}{3} = \dfrac{x}{x+5}$ **b)** $0.4 = \dfrac{m+2}{m}$ **c)** $\cos(34°) = \dfrac{y+1}{y-2}$

Biruni's method relies on being able to measure the height of a mountain in a fairly accurate way. Once students have filled in the appropriate information, they should be able to write:

$$\sin(89.4°) = \frac{r}{r+347.64}$$

The warm-up will prepare them to solve this successfully.

TEACHING IDEA 6:
Research your methods
When students use the Attic stadium measure, they arrive at a circumference of the Earth that is surprisingly close to the actual value. You could ask them to research the method used to determine the actual circumference of the Earth.

TEACHING IDEA 7: The sine rule

INQUIRY   Rather than simply giving students the sine rule, or sine law, let them develop it through the use of right-angled triangle trigonometry.

a) Draw triangle ABC with a right angle at A and label the sides a, b and c.

b) Draw the altitude from A to BC and label it h.

c) Write an equation relating the measures of side a, angle A and altitude h.

d) Write a similar equation relating the measures of side b, angle B and altitude h.

e) Isolate the variable h in each equation.

From here, students should be able to develop the sine rule. Ask them what other ratio might also be equal to the other two.

SUMMATIVE TASK

Measuring the immeasurable

This task could be used as a summative task at the end of a unit on trigonometry. It requires students to apply right-angled triangle trigonometry in a variety of ways and can even include the sine rule if students have learned it (although it is not necessary).

Assessment

If you choose to assess students on this task, you can use criterion D. However, you will have to remove some of the steps, to ensure that students have the opportunity to "select and apply" their own problem-solving techniques. Another option is to remove that strand of the criterion. Just make sure that you assess it at least twice during the school year with other assessment tasks.

The task-specific descriptor in the top band (7–8) should read that the student is able to:

- select and apply three appropriate methods to determine the inaccessible height (Two of the methods must involve trigonometry.)
- correctly determine the height of the object with each method
- justify the degree of accuracy of the calculated heights
- justify whether the calculated heights make sense in the context of the problem.

Stage 1 of the unit planner

Key concept	Related concepts	Global context
Relationships	Measurement Generalization	Scientific and technical innovation
Statement of inquiry		

Generalizing relationships between measurements has allowed humans to apply scientific principles to the world around them.

In this task, students will use a clinometer to measure angles of elevation of the top of an inaccessible object. The materials that are required include scissors, glue, cardboard, string, weights and a photocopy of a protractor. While primitive, this clinometer mimics what more modern clinometers do, which is measure angles of elevation and depression. Students should practise finding the angle of elevation of a variety of objects before attempting the summative task. It is important to note that it is easier if students just read the larger of the two angles on the protractor because 90° must be subtracted from all measurements (since the weight hangs at right angles to the line of sight). For example, when looking out horizontally, the weight hangs at 90° even though the angle of elevation is zero. When looking straight up at 90°, the weight hangs at 180° (or zero).

WEB LINKS

The instructions to make a clinometer can be found by searching for clinometer at http://nrich.maths.org

INTERDISCIPLINARY LINKS

Design

Building a clinometer could be a task to suggest to the design teacher. Students could investigate the problem of measuring angles of elevation accurately and then construct a clinometer to accomplish it.

Once students have constructed their clinometers, let them measure the height of an object of known height, such as the school building or a nearby monument. Hold a friendly competition to see who comes closest to the actual height. You can then discuss sources of error in using the clinometer and degrees of accuracy with taking and recording measurements.

The diagrams for methods 1 and 2 are given below. Method 3 is left up to the students to research, but they may choose methods such as using a mirror and similar triangles.

Method 1

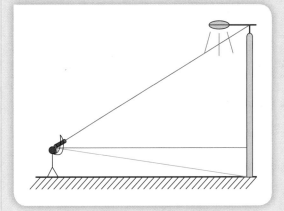

Method 2

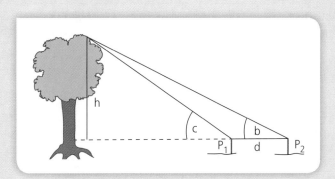

The method used in the Great Trigonometric Survey of India is method 2.

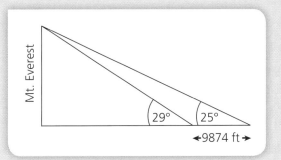

CHAPTER LINKS

The metric relations and/or the formula for multiple right-angled triangles, which are both developed in Chapter 14 on Simplification, may be used for method 3.

Standing over 100 miles away from Mount Everest, the British made two angle measurements from spots a known distance apart. Using trigonometry, they calculated the height of Everest and were within 30 feet of today's accepted value! Since Everest is still growing, they may have been even closer than we think. (Make sure that students practise this method at school.)

Summary

Students often think that measurement is simply something that they do with a ruler. In this chapter, students have investigated what it would take to measure objects and they have used measurements to approximate other measures. Their definition was also expanded as they used these measurements to make conjectures about relationships in circles that were ultimately proven geometrically (circle theorems). Finally, they used known measurements to determine inaccessible ones. Where measurement may sometimes be seen as a result, in this chapter it became a starting point for the acquisition of new knowledge.

10 Model

A depiction of a real-life event using expressions, equations or graphs

	ATL skills	Mathematics skills
TOPIC 1 Linear and quadratic functions		
Activity 1 Images	✓ Apply existing knowledge to generate new ideas, products or processes.	**Algebra** ✓ Find the equation of a line, given two points on the line. 　Optional: Absolute-value functions can be used instead of two distinct linear functions. ✓ Find the equations of linear and quadratic functions, given the graph. ✓ Define a quadratic function in standard and vertex forms. ✓ Describe a piecewise-defined function.
Activity 2 Tunnel shapes	✓ Listen actively to other perspectives and ideas.	**Algebra** ✓ Define a quadratic function in standard form.
TOPIC 2 Regression models		
Activity 3 Going for gold	✓ Identify trends and forecast possibilities.	**Algebra** ✓ Find the equation of a line, given two points on the line. **Statistics and probability** ✓ Find a line of best fit by sight, and using technology.
Activity 4 Dental arches	✓ Make unexpected or unusual connections between objects and/or ideas.	**Algebra** ✓ Define a quadratic function in standard form. **Statistics and probability** ✓ Use quadratic regression to determine an equation to represent a real-life situation.
Activity 5 Dengue fever	✓ Collect and analyse data to identify solutions and make informed decisions.	**Statistics and probability** ✓ Use quadratic and exponential regression to determine an equation to represent a real-life situation.
TOPIC 3 Scale models		
Activity 6 Sprinklers	✓ Use models and simulations to explore complex systems and issues.	**Geometry and trigonometry** ✓ Find the distance between two points and the shortest distance from a point to a line. ✓ Find the area of triangles, circles and sectors of circles.
Activity 7 Urban planning	✓ Structure information in summaries, essays and reports.	**Geometry and trigonometry** ✓ Find the distance between two points and the shortest distance from a point to a line. ✓ Select the most appropriate triangle centre and justify why it is the best choice in a given context.
OTHER RELATED CONCEPTS	Space　　Representation　　Simplification　　Justification	

Introducing model

The aim of teaching mathematical modelling is for students eventually to pose their own questions about the world in which they live, and learn to use mathematics to answer those questions. In this chapter, students will model real-life situations, using different areas of mathematics. Figure 10.1 summarizes the modelling process.

The important modelling skills to highlight in this chapter are:

- the translation of the real-life problem into a mathematical model
- an analysis of the mathematical model, leading to its solution
- an interpretation of the mathematical solution in the context of the real-life problem
- an application of the model to other situations, taking into account its limitations.

Perhaps what we struggle with most in mathematical modelling is finding sources of interesting and mathematically accessible problems. Situations with which students can identify are always good starting points. Once we have established that the mathematics inherent in the problem is accessible to our students, we should encourage the students to identify the key assumptions that they will have to make in order to name the variables, parameters and constraints of the problem. Remind them that they should look at more than one possible model, making them aware that in modelling there is seldom just one correct answer.

Figure 10.1 Mathematical modelling

Without the help of mathematics, art could not advance and all the sciences would perish.

Malba Tahan

TOPIC 1 — Linear and quadratic functions

Confucius, probably one of the most famous philosophers in history, once said that in order fully to understand a process, it is not enough to go from A to B, it is also important to be able to go from B to A. Students are used to graphing functions, particularly using a GDC, but are less used to finding a function from its graph. This reverse process requires practice and understanding of essential function definitions.

In this topic, students will define linear and quadratic equations, given their graphs, and will be asked to express functions as transformations of other functions (including absolute-value functions).

Before they attempt to model the world they live in, you might give students practice in modelling idealized forms and shapes. Let students experiment with easier shapes and forms first, including some they make themselves, before attempting Activity 1 in the student book.

You might suggest that students draw a simple stick figure on graph paper and attempt to find the functions that define the figure. They will see the need for making an appropriate scale, and for using the intersections of the grid lines when drawing lines on the graph, to make it easier to identify points on the line. They could then confirm their functions, using the GDC or graphing software.

This activity can be conducted in a group, or as a peer or self-assessment activity.

Before beginning this activity, students will need to know the syntax for:
- entering functions in their GDCs
- entering restricted domains
- entering a function as a transformation of a previous function.

The first time students attempt this activity, they might enjoy trying it without any prompting from the teacher. To help them get started, hold a discussion on choosing the best form of the quadratic function to work with.

- The vertex form is the most appropriate, when the coordinates of the vertex can be clearly identified.
- The x-intercept or factorized form is best, when the zeros of the function can be clearly identified.

The leading coefficient of the quadratics in both images is 1, except for Snowy's eyes. You could discuss with students how to obtain the leading coefficient either analytically, or graphically by trial and error. Students may need help to understand that they can enter some of the functions as transformations of another function, for example, in the design of Snowy's hat. Also, two piecewise defined functions could be entered as one absolute value function, for example, for Snowy's arms.

Functions for the ghost

Body part	Function	Domain restriction
Head	$y = -(x-4)(x+2)$	$-2 \leqslant x \leqslant 4$
Left eye	$y = 5$	$-0.5 \leqslant x \leqslant 0.5$
Right eye	$y = 5$	$1.5 \leqslant x \leqslant 2.5$
Smirk	$y = (x-1)^2 + 2$	$0.5 \leqslant x \leqslant 2$
Left arm	$y = -(x+4)(x+8)$	$x < -4$
Left sag	$y = (x+2)(x+4)$	$x < -2$
Right arm	$y = -(x-6)(x-10)$	$x > 6$
Right sag	$y = (x-4)(x-6)$	$4 < x < 6$

Functions for Snowy

Body part	Function	Domain restriction
Face	$y = (x-4)^2 + 3$	$3 \leqslant x \leqslant 5$
Body	$y = -(x-4)^2 + 3$	$-0.5 \leqslant x \leqslant 0.5$ and $2 \leqslant x \leqslant 6$
Left eye	$y = -(x-3.5)^2 + 4.7$	$3.3 \leqslant x \leqslant 3.7$
Right eye	$y = -10(x-4.5)^2 + 4.7$	$4.3 \leqslant x \leqslant 4.7$
Smile	$y = (x-4)^2 + 3.3$	$3.8 \leqslant x < 4.2$
Rim of hat on face	$y = -(x-4)^2 + 6$	$3 \leqslant x \leqslant 5$

TIP

If students are not familiar with the transformations necessary to obtain the function $y = a(x-h)^2 + k$ from $y = x^2$, you could let them investigate separately the effects that changing the parameters a, h and k will have on the original quadratic. Students may further investigate the relationships of $f(x) = x^2$ to $y = -f(x)$ (reflection in x-axis), and $y = f(-x)$ (reflection in y-axis).

TIP

Some GDCs, such as the TI-84+, only allow for a maximum of 10 functions to be plotted on the same graph. This will pose a challenge, since students often do not realize that they can enter on one line a function defined over two different domains, for example, the eyes and whiskers in the Cheshire cat, on the next page.

2nd rim on hat	$y = -(x-4)^2 + 7$	$3 < x < 5$
Top rim of hat	$y = -(x-4)^2 + 8$	$3 \leqslant x \leqslant 5$
Arms	$y = |x-4|$	$1 \leqslant x \leqslant 2.7$ and $5.3 \leqslant x \leqslant 7$

Snowy's hair and the vertical outline of Snowy's hat were drawn with the geometry feature of the TI-Nspire GDC. You could ask your students to find the equations of these lines instead.

EXTENSION

Adjusting the difficulty level

Students might also enjoy using some of their own photos or figures from magazines, and attempting to model a real-life face or shape, or even creating their own pictures. These are excellent open-ended tasks, as students can model as many different functions as they choose (or as specified by the teacher), with varying difficulty levels.

If you have done the vertex form of the quadratic with your students, where the leading coefficient is not 1, but either an integer other than 1 or a rational number, you might challenge them to find the functions that create the cat, below.

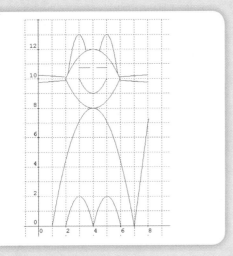

TIP

The Cheshire cat has more than 10 functions, so it is best to use Graphmatica or similar graphing software, or a GDC that can accommodate this number of functions.

Functions for the Cheshire cat

Body part	Function	Domain restriction
Body	$y = -\dfrac{8}{9}(x-4)^2 + 8$	$1 \leqslant x \leqslant 7$
Right leg	$y = -2(x-3)^2 + 2$	$2 \leqslant x \leqslant 4$
Left leg	$y = -2(x-5)^2 + 2$	$4 \leqslant x \leqslant 6$
Head	$y = \dfrac{1}{2}(x-4)^2 + 8$ $y = -\dfrac{1}{2}(x-4)^2 + 12$	$2 \leqslant x \leqslant 6$
Right ear	$y = -4(x-3)^2 + 13$	$2.19 \leqslant x \leqslant 3.53$
Left ear	$y = -4(x-5)^2 + 13$	$4.47 \leqslant x \leqslant 5.81$
Tail	$y = 7x - 49$	$7 \leqslant x \leqslant 8$
Whisker – top right and bottom left	$y = -\dfrac{1}{16}x + \dfrac{41}{4}$ $(y = -0.625x + 10.25)$	$0 \leqslant x \leqslant 2.06$ $5.94 \leqslant x \leqslant 8$

Whisker – bottom right and top left	$y=\dfrac{1}{16}x+\dfrac{39}{4}$ $(y=0.625x+9.75)$	$0 \leqslant x \leqslant 2.06$ $5.94 \leqslant x \leqslant 8$
Eyes	$y=\dfrac{43}{4}$ $(y=10.75)$	$3 \leqslant x \leqslant 3.75$ $4.25 \leqslant x \leqslant 5$
Mouth	$y=(x-4)^2+9$	$3 \leqslant x \leqslant 5$

The intersection points that are not exact points can be found using a GDC.

For the whiskers, you might suggest students draw a line that extends over each whisker. The students will then see that the top right and bottom left whiskers are actually part of the same linear function, as are the bottom right and top left whiskers. Both linear functions pass through the point $(4, 10)$. This will help them to determine the equation of each linear function.

Discovering the equation of a circle

Drawing curves is of course trickier than drawing straight lines, as the only curves that students will have studied are parabolas. When drawing stick figures, students will often want to use circles for faces, and will need help to define the equation of a circle. This would be a good time to lead them to develop the equation of a circle, using the formula for finding the distance between two points.

Ask students, using graph paper, to draw a circle with the centre at the origin and with a radius of 2 units. Then, let the students label any other point on the circle as (x, y), and form an equation in x and y representing the distance from any point (x, y) to the origin. Using the distance formula, they obtain $\sqrt{x^2+y^2}=2$ or $x^2+y^2=4$. Since (x, y) is any point on the circle, they have found the equation of a circle of which the radius is 2. They now only need to solve for y to be able to enter it on the GDC. Since $y=\pm\sqrt{4-x^2}$, they will need to graph separately the two radicals with different signs on the GDC, as shown on this TI-Nspire image.

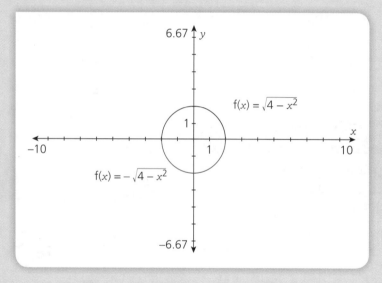

Next, students may develop the general formula for a circle for which the centre is not at the origin by considering vertical and horizontal translations. They may experiment first with expressing the distance between any point (x, y), and another point on a circle with a fixed radius. For example, express the distance between any point on a circle with its centre at $(-2, 1)$ and with radius 2. They should get $\sqrt{(x+2)^2+(y-1)^2}=2$ or $(x+2)^2+(y-1)^2=4$. With a few more examples, they will see that the equation of a circle whose centre is not at the origin is $(x-h)^2+(y-k)^2=r^2$, where (h, k) is the centre of the circle, and r is the radius. This is a translation of a circle with its centre at the origin through $\begin{pmatrix} h \\ k \end{pmatrix}$.

CHAPTER LINKS

In both Chapter 7 on Generalization and Chapter 12 on Quantity, students explore transformations of different graphs. Making explicit links to this throughout this topic will help students see that horizontal and vertical translations result from similar changes in parameters regardless of the original relation.

You might also ask students to attempt to find the functions defining the faces below. They could superimpose their own coordinates over the faces, or they could be given a scale. The TI-Nspire has an animation function that would allow them to see the progression from the sad face to the smiley face.

 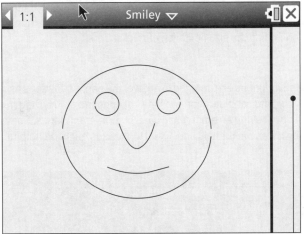

Activity 2 Tunnel shapes

This activity offers students the opportunity to use parabolas to study a real-life problem. Students will design a tunnel that can accommodate large vehicles and, at the same time, they will reflect on the factors they need to consider when deciding the tunnel's shape: safety, environmental impact and, of course, cost of construction. The task itself has two parts: the calculations to answer a list of routine questions about optimal area and a reflection that involves some research about the shape of tunnels.

For the first part, students will benefit from using graphical software or a GDC that allows them to capture pictures, although it is also possible to complete the task without these tools. In this case, students can simply draw the axes on the picture, as shown here, mark the intercept points on the axes and identify their coordinates: V(0, 4) and A(3, 0). The equation of the parabola that models the entrance of the tunnel is obtained easily:

$y = -\dfrac{4}{9}x^2 + 4$. and offers a good opportunity for discussion about the different forms of quadratic equation and when to use each of them.

To find the maximum area of the rectangle that models the cross-section of a big truck, any GDC is suitable: it is sufficient to graph the area function

$y = 2x\left(-\dfrac{4}{9}x^2 + 4\right)$ and find its maximum for $0 < x < 3$. The detail of the domain also offers a good opportunity to talk about restrictions determined by the context rather than by the expression—after all, polynomial functions can be defined in the domain of real numbers but, in this context, the function can just be defined for $0 < x < 3$.

As a group activity, it is also feasible to divide the regions into polygons for which the area formulas are simple, then use symmetry to reduce the number of calculations and estimate this area.

Once students have their model, they could research or be given the wheel-base width of several trucks and calculate the maximum height of truck that could pass through the tunnel safely. This would serve as a good introduction to their work on finding the size limitation of trucks.

EXTENSION

Students might be encouraged to answer the same questions as in the activity, but taking into consideration two lanes of opposite traffic. They might also want to consider having an emergency lane, or multi-traffic lanes. Some research into existing parabolic tunnels and actual truck sizes might be helpful.

TIP

The rest of the task requires some thought. It can form an interesting group activity. Students need to devise a strategy for calculating the areas enclosed by the parabola and the rectangle. Specialist graphing software makes this task easy, as it automatically calculates the areas under curves.

TIP

You could ask students to investigate how varying the location of the axes as well as the form of the quadratic equation that is used to model the tunnel (standard form, vertex form, factored form) affects the ease with which they solve the problem. You might ask them to answer the question, "Are some forms of quadratic equation easier to work with depending on where the origin is placed?" and summarize their findings.

⊂⊃ WEB LINKS

Whispering walls are architecturally intriguing because soft sounds on one end of the construction can be heard far away in another part of the construction. One of the most famous parabolic whispering walls is the Barossa Reservoir dam in Australia. Visit www.panoramicearth.com and search for "Barossa". This website provides a virtual tour of the Barossa, which you might show on an interactive whiteboard. One of the students could be invited to navigate the tour. You might then let students do some research on parabolic whispering walls and explore the mathematics behind their construction.

TEACHING IDEA 1: Everyday parabolic shapes

INQUIRY Activity 2 gives students the opportunity to explore parabolic shapes, such as water fountains, in their everyday lives. Students can take photographs of different parabolic shapes in their own environment and, using graphing software such as Autograph or the software version of the TI-Nspire, transpose the picture onto coordinate axes. The TI-Nspire comes with a variety of pictures that can be used for this purpose. If this technology is not available, superimposing a grid transparency onto their photos will also work.

Instruct students to:
- draw a quadratic function that they think reasonably approximates one of the parabolas in the picture
- use the transformations of functions that they are familiar with to obtain a function that matches one of the parabolic arches
- find the function analytically by selecting three points on the parabola and solving a system of three equations in three unknowns
- use the regression features of the software or GDC to find the quadratic regression function. (This may need to be teacher guided, as quadratic regression is not a topic in the MYP framework.)

This teaching idea also works well with the parabolic dental arch activity in the next topic.

Regression models

Regression analysis can be used to analyse relationships between two variables and could provide insights into that relationship. Modelling these relationships can help us understand scientific and biological phenomena (found in Activities 3, 4 and 5) and help companies design optimal products (found in Activity 4). Calculating the relative strength of this relationship reveals how good the model is as a representation of the real-life situation and how accurate it will be as a tool for prediction.

Activity 3 Going for gold

In this activity, students are provided with the winning times of the men's and women's 100 m races at the Olympics, which follow a fairly linear pattern. Students are asked to determine an equation to model the trend in the winning times of the men's and women's 100 m sprint and to use the model to make predictions on future and past races (extrapolation and interpolation).

In order to discern the pattern, students are first asked to graph the data, using an appropriate scale.

In selecting an appropriate scale, students should first identify the independent and dependent variables. If necessary, remind them that the independent variable is shown on the horizontal axis (x-axis) and the dependent variable on the vertical axis (y-axis). This would be a good time to reinforce the selection of variables as well as graphing conventions. The axes can then be renamed to indicate what is being measured. In choosing an appropriate scale, students should be reminded to consider these points.

- What is the range of results that need to be plotted on each axis? The data should fill the entire graph on both axes, avoiding a cluster of points that are hard to read or leaving a lot of dead space.
- The scale need not be the same on each axis.
- The scale need not start at zero.
- The graph may need to be redrafted if the scale chosen initially does not accurately represent the data.

EXTENSION

Pearson correlation coefficient

The justification of which model is most appropriate usually relies on a statistical measure of some kind. The Pearson correlation coefficient (or the Pearson product moment correlation coefficient, represented by r) can be used to describe the strength and direction of the linear relationship between two variables.

Let students research the Pearson correlation coefficient and answer these questions.

a) Who was Pearson?

b) What does an r value close to 1 or -1 mean?

c) If the value of r is close to 1 or -1, does this mean that a change in one variable caused a change in the other? Explain.

TIP

Although each of the functions is linear, determining whether or not the women will ever run as fast as the men requires solving a system of equations. Be sure to ask students to reflect on their results because the time for the race where men and women tie would be negative!

WEB LINKS

There is an excellent video demonstrating how the times in the men's 100 m race have been changing since the Olympics in 1896. It can be found by searching for "New York Times Every Medalist Ever". There is a similar video for the 100 m freestyle race as well as the long jump.

TIP

Students can find the line of best fit in a variety of ways, depending on the content they have learned. The most rudimentary way is to draw the graph by hand, obtain the mean of the data and plot this point, and then draw the line of best fit by eye, as the line that passes through the middle of the data.

Activity 4 — Dental arches

In this activity, students attempt to determine an equation to model their dental arches, using data that they collect. They are then asked to use the class data to design a mouth guard for an "average" person.

When the students are creating their own teeth impressions and transposing them onto a sheet of graph paper, be sure to discuss the importance of using the same points of reference when placing the axes, as students will be collaborating on data.

EXTENSION

Transformation of the parabola

Ask students to:

- take the original indent points of their individual impression and move the axes so that the origin is at the bottom left indent
- determine the new points on their graph and derive the new equation for the parabolic arch
- state how the two equations compare
- use their knowledge of transformations of functions to explain these differences
- calculate the length and height of the dental arch
- explain their results in comparison to the original equation.

Ellipses

Sometimes a person's dental arch is better modelled by an ellipse. Ask students to find an elliptical equation that fits the individual and average data and comment on its fit in comparison to that of a parabola.

> **CHAPTER LINKS**
> In Chapter 6 on Equivalence there are tasks looking at the different forms of quadratic functions and learning when to use each form, depending on the information provided and the context of the question.

> **DP LINKS**
> In the statistics units in the DP, students will use more advanced tools and processes in regression analysis to improve their ability to analyse statistical models further.

> **CHAPTER LINKS**
> When determining how good a fit the parabola is, use the correlation coefficient of determination. At this grade level, you can simply use a GDC to find the value. If this is the first time students are doing a regression type other than a linear regression, be sure to go through the process of finding the best regression function to represent a model. There is an investigation on this in Chapter 8 on Justification in this teacher book.

 ## Activity 5 — Dengue fever

Assessment

If you choose to assess students on this task, you can use criterion D. The task-specific descriptor in the top band (7–8) should read that the student is able to:

- identify the relevant elements of the dengue fever problem
- select appropriate mathematical strategies to model the spread of dengue fever, and the results of one method used to try to prevent it, accurately
- apply the selected mathematical strategies to reach correct predictions of its potential spread and possible control
- justify the degree of accuracy of the models and resulting predictions
- justify whether the models and predictions make sense in the context of the dengue fever problem.

Key concept	Related concepts	Global context
Relationships	Pattern Change Model	Globalization and sustainability
Statement of inquiry		
Modelling patterns of change can help analyze and predict potential global health hazards.		

In this activity, students are introduced to dengue fever and its growing impact worldwide. They use functions to model its spread as well as to represent scientists' attempts to control the mosquito population that spreads the disease.

Encourage students to graph the data, using either a GDC or graphing software. Students can even draw the graphs by hand as long as they know how to find equations of exponential functions, given points on its curve. Encourage students to consolidate what they have learned thus far and perform a linear and a quadratic regression analysis first. They can then attempt other types of function, based on what has been covered in class.

The increase in the number of dengue fever cases is an example of exponential growth, whereas, once the genetically modified mosquitoes are introduced, the decrease in the mosquito population is an example of exponential decay. You might discuss with students why exponential functions model the data better than the linear and quadratic functions they tried first. How can they tell when a relationship is exponential? The reflection on whether the population can be completely eradicated, both theoretically and experimentally, is especially important as students evaluate the limitations of any model.

TEACHING IDEA 2: Picture mathematics

INQUIRY The study of viruses and bacteria are natural applications of exponential functions. However, the Eiffel Tower presents a unique opportunity for students because its shape has been studied at length and was only recently determined to be exponential. Students could study the Eiffel Tower before attempting Activity 5, as an introduction to exponential regression.

Let students transpose a picture of the Eiffel Tower onto a grid and determine several points on one of its sides (the curve). Then let them use a GDC to find a model for the curve. In reality, the side of the Eiffel Tower is best represented by two exponential functions, one for its upper portion and one for the lower. Students could be asked to find these two different functions, to define the domain of each and to justify their choices.

Extended mathematics You might want students to explore curve-fitting with exponential and logarithmic functions.

TOPIC 3 Scale models

A scale model is a simplification of a real-life situation and is sometimes easier to understand as the variables can be controlled to a greater extent than in real life. We use scale models to test possible outcomes, with the intention of making a decision in an efficient and cost-effective way. For this section, you could start by discussing something simple, such as how to determine the best configuration for furniture in a room and find what pieces will fit, as a basic scale model.

Activity 6 Sprinklers

Assessment

If you choose to assess students on this task, you can use criterion D. The task-specific descriptor in the top band (7–8) should read that the student is able to:

- create a 2D model **to scale** that is **easy to follow** and can be used effectively to help visualize and solve the task
- create the **optimal** configuration for the sprinkler heads
- use a variety of mathematics, including trigonometry and geometry, to **accurately calculate** the percentage of crop land not covered by the sprinklers, the area of land within the configuration to be irrigated and the minimum cost of the pipeline
- **explain** the degree of accuracy used in calculations, given the context of the task
- **critically reflect** on the optimal configuration, given the context of the problem and its suitability in real life.

If you would like to include the reflection component of determining the best alternate configuration that requires no gaps as part of the summative task, you could adapt the rubric to reflect this. It could be used as a way to distinguish between a level 7 and level 8 within the top descriptor level; the reflection component and new configuration would be necessary to achieve a level 8, in addition to all of the correct mathematics required for the other questions.

You could alter the last descriptor point to read:

- *(for a level 8, in addition to the preceding points)* **critically reflect** on the optimal configuration, given the context of the problem and its suitability in real life, and include a scale model of the best alternate configuration based on his/her reflection.

Stage 1 of the unit planner

Key concept	Related concepts	Global context
Logic	Model Space	Fairness and development
Statement of inquiry		
Modelling real-life situations to determine optimal outcomes in an efficient and cost-effective way minimizes wastage of finite resources.		

The mathematical concepts covered are scale and proportions, analytic geometry and trigonometry, so it is important that all students have an understanding of these before they attempt these tasks.

STEP 1 Optimal configuration

Encourage the students to practise their skills in rescaling objects, by working in groups to construct a simple scale model of the Sun and Earth. Give them a sphere to represent the Sun, and ask them to determine the size and orbit of Earth in relation to that sphere. The sphere could be a basketball that you hold up in class (or in the school gym or field, as open spaces are good for this activity) or a famous spherical structure or monument they will have to research.

Students will then have to think about what information they need in order to conduct this activity. As that is part of the modelling process, give them as little information as possible (as long as they have access to resources that will enable them to find the measurements they need).

WEB LINKS

MythBusters is a TV programme created for the Discovery Channel. The show uses elements of the scientific method to test the validity of rumours, myths and news stories. You can see some of the scale models they have created by searching for "mythbusters scale model". Select a video clip of one of the scale models they have created and watch it with the class. Students can then discuss the advantages and disadvantages of the use of scale models.

CHAPTER LINKS

The balloon task in Chapter 9 on measurement can be done by completing a scale model.

TIP

Dynamic geometry software makes a task like this one much easier and is an efficient use of time. An added bonus is that students can verify all of their answers.

Make sure that students know that the task requires them:
- to use basic 2D geometry to determine the percentage of cropland not covered by the sprinklers
- to use trigonometry (cosine law and area of triangle given side lengths) and geometry (area of a sector) to determine the area, within the boundaries of the sprinklers, that is not watered
- to use analytic geometry (shortest distance from a point to a line) to determine the piping cost to install the irrigation system.

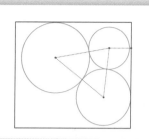

The configuration students will come up with looks similar to the illustration shown on the right (if the pipe runs along the right-hand side of the crop field).

Since this task would be done as a summative task at the end of a trigonometry or analytic geometry unit, students would already know the cosine law and so should be able to find the angles without difficulty. Students should also be able to find the formula for the area of a sector, but if they need a hint you could tell them to look at the ratio of the sector angle in comparison to a whole circle. Students struggle most with finding the area of the triangle if they do not already know the formula. The quick think in the student book can help students derive the formula they need.

TIP

If they know it, students could use Heron's formula:

$A = \sqrt{s(s-a)(s-b)(s-c)}$ where a, b and c are the lengths of the sides of the triangle and $s = \dfrac{1}{2}(a+b+c)$.

To determine the length of piping used, students need to find the shortest distance from a point (one triangle vertex) to a line (piping running along the edge). You can do a demonstration showing that this occurs when the original and a created line that passes through the point are perpendicular to each other. If students have studied analytic geometry they may have covered

how to do this, but if not, you can give them the formula $d = \dfrac{|Ar+Bs+C|}{\sqrt{A^2+B^2}}$, for a line given by the equation $Ax+By+C=0$

and a point (r, s).

REFLECTION

One of the reflection questions asks students to consider what to do about the sections of the field that will not be covered by the sprinklers. In reality, the likelihood is that the sprinklers would be set up with overlap as the effort required to water the section not covered would take priority over the water wasted by the overlap. Based on this idea, you could ask students to determine the best way to overlap the sprinklers to minimize water wastage. The amount of scaffolding you give them will depend on the age group and the time you allow for the investigation. If you were including this as part of a summative task, you would limit the scaffolding to ensure students are selecting and applying their own problem-solving strategies.

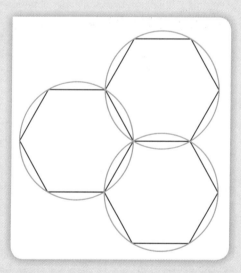

To simplify the model you need only look at three sprinklers (see the diagram on the right).
The following hints could be given to students.
- We have simplified the model by looking at three sprinklers, but we may want to add more into the pattern to cover a larger area—would it be easier to use the same size sprinkler range?
- How do you ensure no gaps—do you remember your rules regarding tessellations?
- How can we incorporate a shape that tessellates with a circle?
- What is the largest regular polygon that can tessellate in a rotation (although you may want them to investigate all regular polygons that can tessellate in a rotation and test the equilateral triangle and the square to compare to the regular hexagon)?
- What would the overlap be with two sprinklers? Now add a sprinkler and see what the overlap will be when one more is added. Will this pattern continue for each additional sprinkler?

TIP

If you want to change the topic you could make the application internet wireless coverage in an office or coffee shop.

Activity 7 — Urban planning

As this task is open ended you should expect to see a variety of ideas and different community layouts, but students must include certain essential infrastructure for it to be a workable community. The focus should be on the students' justifications for their location choices, based on clear explanations of the different analytic geometry concepts taught, such as triangle centres and shortest distance from a point to a line.

Essential infrastructure with possible locations might include:

- City centre (shops and businesses)—at the centroid
- Hospital/medical centre—should be located at the circumcentre, which is the point where it is the same distance from the three residential housing estates
- Fire station—located between the circumcentre and the centroid, so that it is almost the same distance from each of the three residential housing estates and is near the city centre
- Police stations—perhaps two police stations to serve the area, one placed between Residential Housing Estate A and Residential Housing Estate B and the other between the central area of the city and Residential Housing Estate C
- High school—one near the centroid
- Primary schools/daycares—three primary schools located within the three residential housing estates
- Major roads—these are essential in a city because they allow for easy commuting and lessen traffic jams. They must follow a logical structure
- Railway station—on the railway line closest to the centroid because it should be near the city centre (shortest distance from a point to a line).

Students might include other key infrastructure students such as: professional and government buildings, banks, a public library , a community centre/sporting complex, religious centres, art galleries and other cultural buildings, clinics—doctors, dentists and other specialists, pharmacies, supermarkets, petrol stations, restaurants and shops (retail and professional), a cinema and parks and recreation areas (perhaps even an artificial lake). They might also include a cemetery (far from the residential housing estates and the city centre).

A discussion of utilities is essential but does not need to be included in the urban plan as the detail required to put in these grid systems would be onerous. Students should mention sewerage, electricity, gas, water, telephone/internet/tv cable lines. Public transport is a key service for any community and there should, at least, be buses to provide citizens with inexpensive transportation that connects to the railway service. Mention of what public services need to be offered such as solid waste management is also important.

Assessment

If you choose to assess students on this task, you can use criterion D. The task-specific descriptor in the top band (7–8) should read that the student is able to:

- create a 2D model **to scale** that is colour-coded appropriately to make it **easy to follow**
- include **all** essential infrastructure **positioned correctly** with **excellent justification** for chosen location through clear explanations of what mathematical concepts were used to determine them
- **accurately calculate** the distances between **all** essential infrastructure in reference to each other and the three housing estates
- include **most** key infrastructure and **justify** their location

TIP

This task could be used in place of the sprinkler task for classes that have not yet studied trigonometry.

TIP

The coordinates and line are arbitrary in this activity so you can change them to suit your means or even give the students a map of land that is to be developed into housing estates and get them to use that as a basis for their housing estate locations, transportation routes and scale.

WEB LINKS

An interesting way to get students thinking about what services and infrastructure to include is to play a game such as "SIMCITY". Free downloads of the basic game are available.

- discuss **most** of the other necessary infrastructure needed to make a workable community
- **explain** the degree of accuracy used in calculations given the context
- show a real depth of understanding of the project and excellent creativity and originality in planning the community.

Stage 1 of the unit planner

Key concept	Related concepts	Global context
Logic	Model Space	Globalization and sustainability
Statement of inquiry		
Modelling real-life situations to determine the optimal location of resources ensures the needs of communities are addressed in an efficient and cost-effective way.		

Summary

Through the activities in this chapter, students have encountered a wide variety of real-life problems or situations, as well as a diverse range of mathematical areas to model them. We hope this chapter will encourage students to consider the mathematics behind their own areas of interest or practical everyday problems. A possible follow-up to this chapter is to assign mathematical explorations to your class, where the students themselves select areas of interest in their everyday lives to model mathematically. They might want to do this individually or with another student. Having to present their explorations to their class will help to reinforce the need for clear and correct language and mathematical communication. Ideally, it will also serve to confirm the reality that indeed mathematics is everywhere!

 **DP LINKS**

The DP requires that students complete a mathematical exploration (HL and SL), or mathematical project (Mathematical Studies), as part of their final IB mathematics grade.

11 Pattern

A set of numbers or objects that follow a specific order or rule

	ATL skills	Mathematics skills
TOPIC 1 Famous number patterns		
Activity 1 Triangular numbers	✓ Understand and use mathematical notation.	**Geometry and trigonometry** ✓ Use the properties of similar shapes to determine patterns.
Activity 2 Polygonal numbers	✓ Understand and use mathematical notation.	**Algebra** ✓ Determine patterns in recursive functions.
Activity 3 Square and cubic routes	✓ Propose and evaluate a variety of solutions.	**Algebra** ✓ Find and justify general rules.
Activity 4 The binomial theorem	✓ Propose and evaluate a variety of solutions.	**Algebra** ✓ Find and justify general rules for formulae.
Activity 5 Solving endless equations	✓ Consider multiple alternatives, including those that might be unlikely or impossible.	**Algebra** ✓ Use the quadratic formula and factorizing to solve quadratic equations. ✓ Expand and simplify algebraic expressions.
TOPIC 2 Algebraic patterns		
Activity 6 Investigating squares	✓ Encourage others to contribute.	**Number** ✓ Calculate absolute value. **Algebra** ✓ Factorize algebraic expressions.
Activity 7 Investigating trinomials	✓ Encourage others to contribute.	**Number** ✓ Find the common factor of two integers. **Algebra** ✓ Factorize algebraic expressions.

Activity 8 Mosaic tile patterns	✓ Propose and evaluate a variety of solutions.	**Algebra** ✓ Determine the general rule and function that represents linear and quadratic patterns.
Activity 9 The locker problem	✓ Practise visible thinking strategies and techniques.	**Number** ✓ Find the common factor of two integers. **Algebra** ✓ Find and justify general rules.
Activity 10 Exponential growth and decay	✓ Make connections between subject groups and disciplines.	**Algebra** ✓ Determine the general rule and function that represents exponential patterns. ✓ Define the parameters of exponential functions.
Activity 11 The Titius–Bode law	✓ Consider ideas from multiple perspectives.	**Algebra** ✓ Find and justify general rules. ✓ Define an arithmetic sequence with an explicit formula and a recursive formula. ✓ Define a number sequence with a general formula.
Activity 12 The chaos game	✓ Consider ideas from multiple perspectives.	**Geometry and trigonometry** ✓ Use the properties of similar shapes to determine patterns.
OTHER RELATED CONCEPTS	Generalization Representation Justification Model	

Introducing pattern

Pattern is one of the most prevalent themes in mathematics. At this point in their education, students have been exposed to a rich array of patterns in different areas of mathematics. They will now have an opportunity to revisit some famous patterns they may have already seen, such as Pascal's triangle, the Fibonacci sequence, and triangular and other polygonal numbers. The subsequent activities invite students to solve problems by discovering underlying patterns.

Although fractals (covered in Chapter 15) and chaos theory (Chapter 11), as topics, are not a part of the MYP subject framework, it is important to introduce students to these new and exciting branches of mathematics.

> *Though the structures and patterns of mathematics reflect the structure of, and resonate in, the human mind every bit as much as do the structures and patterns of music, human beings have developed no mathematical equivalent to a pair of ears. Mathematics can only be "seen" with the "eyes of the mind". It is as if we had no sense of hearing, so that only someone able to sight-read music would be able to appreciate its patterns and harmonies.*
>
> Keith Devlin, *Mathematics: the Science of Patterns*

Number patterns have always fascinated mathematicians. They provide an interesting path by which to engage students in this chapter.

Activity 1   Triangular numbers

Students will explore many interesting properties of triangular numbers in this chapter. As students are discovering the pattern, the investigation in this activity is a good starting point. They use a graphical and visual method to derive the formula, by finding the area of the rectangle and dividing by two to get the area of the "triangle".

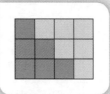

For the third row of the pattern, they have shaded six squares so when they invert the pattern and put it next to the shaded squares, they make a rectangle with a height of three squares and a width of four squares. Ultimately, they reach the formula $\frac{n(n+1)}{2}$, where n is the position of the number in the sequence.

TEACHING IDEA 1: High-five!

As an alternative to Activity 1, you could try this one. It produces the same results, but gets students moving.

Ask: "How many 'high-five' greetings will take place, if every student in the class high-fives every other student just once?" Encourage students to try it out. They could make a table to record their results.

Number of students	T_n – the total number of high fives
1	1
2	3
3	
4	
5	
6	
7	
8	
9	
10	
30	
100	
Any number of students, n	

TIP

In the student book the students are asked to reflect on why they think that the sequence of numbers they found is called the triangular numbers. You may want to give them the hint to draw the numbers as dots instead of shaded squares. If they draw them instead as dots, then each of the numbers makes a triangle.

TIP

It starts to get difficult to record the number of high-fives when you have four students or more, so ask students for some strategies on how best to represent this information in a diagram.

Ask students to investigate these scenarios.

- How many high-fives occur when a group of n people high-five exactly once with every other person in the group? Derive the mathematical formula that can be used to determine the number of high-fives in any given group of people.

- If there are 40 people in a group, how many high-fives will take place?

- At a party, everyone gave everyone else a high-five. There were 100 high-fives altogether. How many people were there at the party?

You could also do both activities, to demonstrate how two approaches can lead to the same pattern and conclusion.

After completing Activity 1, there are a few possible extension activities involving triangular numbers and other polygonal numbers.

TEACHING IDEA 2: Pascal's triangle

You could ask students these extension questions about triangular numbers.

a) Look at Pascal's triangle. What connections can you see between triangular numbers and the triangle?

b) Is there a way to determine the sum of the numbers in the sequence of the pattern in Activity 1 just by looking at the triangle?

c) Look at your table of triangular numbers, and write it out as a sequence of numbers. What do you notice about the sum of any two adjacent triangular numbers? Do you think this is always true?

d) There are two ways that you can attempt to prove your conjecture, geometrically and algebraically.

    (i) Geometrically: draw the triangular numbers with dots to form a growing triangle. Think about the shape you get when you put together any two adjacent triangular numbers. Explain.

    (ii) To prove your conjecture algebraically, you could first write out each triangular number as an arithmetic sequence with a common difference of one. Then, using the formula for the sum of an arithmetic sequence, write each of the two adjacent numbers in general form, and simplify. Write and interpret the result you get.

> **EXTENSION**
>
> An interesting point is that the sum of the reciprocals of all the triangular numbers converge on, or get closer and closer to, two. Can you explain why this is the case? The trick is to write the reciprocal of a triangular number sequence as a difference of fractions.

Once students have completed Activity 1 on triangular numbers, they are ready to explore the family of numbers this number pattern belongs to–polygonal numbers.

 Activity 2   **Polygonal numbers**

As this activity includes many formulae, it is worth exploring them further with some numerical questions.

    a) Is 300 a triangular number? If so, state its position in the sequence.

    b) Show that 2016 is a hexagonal number and state its order.

    c) Find a list of numbers that are both triangular and pentagonal.

> **TIP**
>
> This activity works better if students use spreadsheets to generate several examples of polygonal numbers, as this will produce enough data to spot the pattern.

WEB LINKS

Visit http://demonstrations.wolfram.com. This website offers free downloads of interesting interactive mathematics applications.

Activity 3 **Square and cubic routes**

In this activity, students have to find the number of shortest routes between two points on a grid. Begin by asking them to think about what "shortest route" means in this context. Once they understand this, they just need to move either up or to the right. It is important to guide them towards the use of simple notation such as, for example, U = 1 square up and R = 1 square to the right. With this notation it is easy and quick to list all shortest routes for step 1: UURR, URUR, URRU, RUUR, RURU and RRUU. You can also print a grid and let the students draw all possible routes. This may motivate them to adopt a quicker method, as the number of routes follows the Pascal triangle pattern and the values increase quickly. Here are the answers to the steps.

STEP 1 6 or $\binom{4}{2}$ as, taking A as the top vertex of the Pascal triangle, B is located on the fourth row, position 2 or "from the 4 segments select 2 when you go right (or up)".

STEP 2 20 or $\binom{6}{3}$ as from the six segments you need to select the three where you go right or up.

STEP 3 Taking A as the top vertex of the Pascal triangle, from A the number of ways to each point is given by the corresponding Pascal triangle entry, so to get to each vertex of the square you need to add the values of the adjacent vertices.

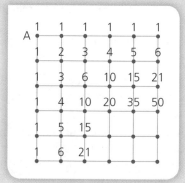

STEP 4 It is better if students calculate the number of routes from A to C and from C to B, and then discuss what to do with these values. Again, the use of diagrams and simple notation, as shown above, may help them to realize that for each of the six routes from A to C they have two possible ways of getting to B. It is worth asking questions similar to this one, in case students do not notice that they need to multiply the number of ways to get to C and to B.

EXTENSION

The extension task is difficult because it involves adding more routes. A model of the cube skeleton, made from straws, may make this task accessible to most students, as they can just use stickers to indicate the number of routes to each vertex and keep adding the values of adjacent vertices to obtain the next value.

As an additional extension, you could ask students to find a general solution to the problem.

TEACHING IDEA 3:
Research famous patterns
To bring together activities 1, 2 and 3, ask students to do some research about the Pascal triangle and polygonal numbers and to extend this to tetrahedral and cube numbers. It may also be interesting to research the history of polygonal numbers, including Fermat's conjecture on polygonal numbers that has challenged many mathematicians for a long time.

In this inquiry-based activity, students explore the binomial theorem and arrive at a method for multiplying binomials that uses Pascal's triangle. In the end they should be able to find the coefficient of any term in a binomial expansion.

The expansion of $(x+a)^n$ carries the added difficulty of locating the values from Pascal's triangle within the coefficients.

TEACHING IDEA 4: Patterns in binomial expansion

INQUIRY Before moving on to question **e)**, ask students to expand these expressions

a) $(x+a)^2$ **b)** $(x+a)^3$ **c)** $(x+a)^4$

Ask them to look for patterns in the exponents of both x and a along with the values from Pascal's triangle. Asking them to fill in x^0 and a^0 may make future work easier, for example:

$$(x+a)^4 = 1x^4a^0 + 4x^3a^1 + 6x^2a^2 + 4x^1a^3 + 1x^0a^4$$

Students should see that the exponent of x decreases in one direction, while the exponent of a increases. However, the degree of each term is always the same. They should notice the symmetry of the coefficients and exponents.

Let them now begin question **e)**.

Provide those who struggle with the pattern in question **f)** with this structure.

	Expanded form	Factor out the appropriate value from Pascal's triangle
$(x+3)^2$	$x^2 + 6x + 9$	$1(x^2) + 2(3x) + 1(9)$
$(x+3)^3$	$x^3 + 9x^2 + 27x + 27$	$1(x^3) + 3(3x^2) + 3(9x) + 1(27)$

Encourage students to look for the same patterns in the exponents here as they found in the above activity with $(x+a)^n$. This will require them to rewrite 9 as 3^2 and 27 as 3^3. Expanding a binomial, then, follows the same pattern as $(x+a)^n$ but with the values of Pascal's triangle inserted appropriately. The solution to question **g)** should now be easier.

TIP

As students work through the activity, the most typical mistakes will be in multiplying/expanding the binomials. Because expanding with large exponents can be tedious, encourage students to use the result from $(x+y)^2$ as the starting point for the expansion of $(x+y)^3$ rather than starting again from the beginning. They can then be asked to expand $(x+y)^4$ by squaring the result of $(x+y)^2$ and verifying their answer with the product of $(x+y)$ and the result of $(x+y)^3$.

For students to be able to answer question **j)**, they will need to have a firm understanding of using the binomial theorem when expanding expressions. It would be worthwhile to practise as many times as necessary, before attempting to find the coefficient of a single term where they need not expand entirely. If they understand and can verbalize the pattern, they should be able to see that the requested term in part **(i)** should include $(2x)^2(-3y)^4$ along with the appropriate value from Pascal's triangle (6). The coefficient is then $6 \times 4 \times 81$, which is 1944.

TEACHING IDEA 5: Other ways of finding the value from Pascal's triangle

There are several ways of finding the appropriate value from Pascal's triangle without recreating the triangle. One method is to look at each term individually and answer similar questions.

$(x+a)^4 = 1x^4a^0 + 4x^3a^1 + 6x^2a^2 + 4x^1a^3 + 1x^0a^4$

The coefficient of x^3a^1 is the number of different ways of arranging three xs and one a:

$xxxa \quad xxax \quad xaxx \quad axxx \ldots$ (4 ways)

Similarly, the coefficient of x^2a^2 is the number of ways of arranging $2x$s and $2a$s:

$xxaa \quad xaxa \quad axxa \quad xaax \quad axax \quad aaxx$ (6 ways)

Though lengthy, this can be used for any of the coefficients and may be shorter than recreating the triangle when students are looking for the coefficient of just one term.

EXTENSION

Combinatorics: The above teaching idea demonstrates how the values in Pascal's triangle are related to a branch of mathematics called Combinatorics, which involves permutations and combinations (arrangements of items). If students will be learning this topic, or are interested in another way of determining the elements of Pascal's triangle, they should discover that the values in the triangle:

```
              1
           1     1
        1     2     1
     1     3     3     1
   1     4     6     4     1
 1    5    10    10    5    1
```

are actually the same as the values obtained from the following calculations (used in problems involving combinations):

$$\begin{array}{ccccccccc}
& & & & \binom{0}{0} & & & & \\
& & & \binom{1}{0} & & \binom{1}{1} & & & \\
& & \binom{2}{0} & & \binom{2}{1} & & \binom{2}{2} & & \\
& \binom{3}{0} & & \binom{3}{1} & & \binom{3}{2} & & \binom{3}{3} & \\
\binom{4}{0} & & \binom{4}{1} & & \binom{4}{2} & & \binom{4}{3} & & \binom{4}{4}
\end{array}$$

TIP

The notation $\binom{n}{r}$ is the same as $_nC_r$, which means the number of combinations of n objects taken r at a time.

Students should then be prompted to develop a formula for all of the terms in a binomial expansion, using notation such as: $(a+b)^n = \sum_{k=0}^{n} \binom{n}{k} a^{n-k} b^k$

Activity 5 — Solving endless equations

In this activity, students learn ways of solving endless equations by first making a simple substitution. They then solve the resulting quadratic equation. Encourage students to use different techniques, for example, the quadratic formula, or completing the square. Discourage the use of a graphical method, as the exact answer is not accessible. It is important that students leave the answers exact, to see the emerging pattern of phi. This leads to the definition of the golden rectangle, and to the discovery of other numbers that can be written as endless equations. The activity ends with an assignment on the natural constant e, which can also be written as an endless equation.

Due to the connections between Fibonacci numbers and the golden ratio, the golden rectangle can be approximated via a sequence of rectangles with dimensions that are the Fibonacci numbers, as here. You may want students to explore these connections.

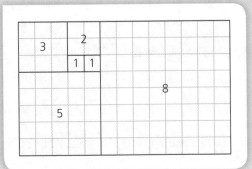

TOPIC 2 — Algebraic patterns

In this topic, students discover some relationships between numerical and algebraic factorization.

Activities 6 and 7 — Investigating squares and Investigating trinomials

For both activities in this topic, students will need to be familiar with factorizing the difference between two squares, and trinomial factorization. If they haven't yet learned this, the following teaching idea will help them discover the factorization rules.

It is easier for students to determine the steps involved in factorizing a trinomial expression if they see a visual representation of how to factorize, using algebra tiles. You can either use actual algebra tile manipulatives or you can let the students use virtual manipulatives.

⊙ WEB LINKS

The website http://nlvm.usu.edu has many excellent virtual manipulatives that you can use for student inquiry.

TEACHING IDEA 6: Using algebra tiles to determine the steps to factor algebraic expressions

For each trinomial, show how you arranged your algebra tiles to determine the final product. For example:

$$x^2 + 7x + 10 \text{ has factors } (x+2)(x+5)$$

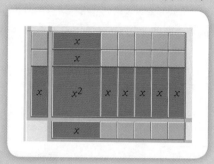

For each of the following trinomials, draw the rectangle algebra tiles diagram and list the factors.

$x^2 + 3x + 2$ $x^2 + 5x + 6$ $x^2 + 8x + 12$

$x^2 + 5x + 4$ $x^2 + 7x + 10$ $x^2 + 6x + 9$

a) Can you see a pattern developing between the trinomial and its factors?

b) Look at the factors of these trinomials and then look at the middle term and the end term of the trinomial. What is the relationship?

c) Given this relationship, write down the general steps that you would use to factorize these types of trinomial.

d) Can your general steps be used for trinomials containing negative integers?

Try factorizing these, to find out.

   (i) $x^2 + 3x - 4$ **(ii)** $x^2 - 6x + 8$ **(iii)** $x^2 - 4x - 12$

Do your steps also work for special cases? How would you write these binomials as trinomials with the same form as the examples above?

e) Try factorizing these.

   (i) $x^2 - 16$ **(ii)** $x^2 - 1$

Why do you think these are considered special cases?

f) For the following questions, you will need to think about what happens when you combine factorizing of numbers with factorizing trinomials.

   (i) $4x^2 + 16x - 48$ **(ii)** $-2x^2 + 18x - 16$ **(iii)** $3x^2 - 27$

In both activities in this topic, students can consolidate their factorization skills. Take care with the mathematical notation and terminology. In activity 6, students should arrive at the conjecture: for any two integers a and b, such that $a > b, |a^2 - b^2| = d(a+b)$, where $d = a - b$. The justification for this conjecture is the rule for factorizing the difference between two squares. Since $a^2 - b^2 = (a-b)(a+b)$, and $a - b = d$, then the result follows.

In activity 7, students will discover some patterns that are helpful in factorizing trinomials with large coefficients. Students can easily factorize the first three trinomials, and also find the other three values of b that satisfy the stated condition: $b = 16, 17,$ and 61. For part **c)**, students should discuss the relationship of b with regard to 6 and 10 to explain that the factors would be grouped in pairs to produce b, but would only be used once to avoid repetition, hence there would be six values for b. Recognizing both the pattern and rule for factorizing trinomials, they will know that interchanging a and c will have no effect on possible values of b. For **f)**, students should use the fact that $12 \times 15 = 180$, 180 has 18 factors, hence there will be nine possible values for b. Students may need some help with their notation and terminology in summarizing and generalizing their findings.

Applying number patterns

In this topic, students look at naturally occurring and man-made patterns and determine and describe these patterns in different real-life contexts.

 Activity 8 **Mosaic tile patterns**

Students can find it difficult to derive an algebraic equation from a table of values, so it is important that they have a systematic approach when finding linear and quadratic rules. The first step is to find the first and second differences to determine the type of function they are dealing with. In this activity, they will determine the basic linear and quadratic patterns in a tile design. They will do this without the aid of technology.

The sequences for the green and purple tiles are linear and the patterns are relatively easy to find, but the orange tiles follow a quadratic rule that is more difficult to identify. You could give them a hint: tell them to find the number of orange tiles in the "zero" stage. This will give them the value of c in the general equation $y = ax^2 + bx + c$, where x represents the stage number and y represents the number of tiles. Then they can pick two of the ordered pairs, set up two equations and solve them simultaneously to find the values of a and b. They can use the same process to find the quadratic expression to represent the total number of tiles, but the most efficient way is simply to add the three algebraic expressions together.

Purple tiles	$y = 4x + 8$
Green tiles	$y = 8x + 8$
Orange tiles	$y = 4x^2 + 4x$
Total tiles	$y = 4x^2 + 16x + 16$

Here, x represents the stage number and y represents the number of tiles.

Once the students have found all the algebraic rules, they can put the data into a GDC and verify their solutions.

TIP

To find the pattern, look at the differences between pairs of consecutive numbers in the sequence. For example, the purple tiles have the sequence 12, 16, 20, 24, 28, so when you subtract one number from the number before it, you get 4 each time. Since the differences all have the same value, you know it is a linear relationship. For the orange tiles you have the sequence 8, 24, 48, 80, 120, so the first differences are 16, 24, 32, 40. They are clearly not the same. Continue the process and find the second differences of 8. Since these are all the same, then the relationship is quadratic.

EXTENSION

a) Which colour of tiles is growing:
   (i) at the slowest rate
   (ii) at the fastest rate?

b) How many tiles of each colour will be in:
   (i) the 6th stage of the design
   (ii) the 10th stage of the design
   (iii) the zero stage? (Show where this would be on the original diagram.)

c) At what stage number will the design:
   (i) use 68 purple tiles
   (ii) use 624 orange tiles
   (iii) use 1600 tiles in total?

Activity 9 The locker problem

In this activity, students will work on the classic locker problem. They will discover and use real number properties in order to justify their results. It is very important that, before they begin, every student understands the instructions. It would be worthwhile taking a bit of time to go through, for example, the first 10 athletes' opening and shutting of the lockers. Although going all the way up to the 30th athlete is required in question **c)**, once they spot the pattern, they no longer have to go beyond this number for individual cases.

Students should discover these criteria.

- Each athlete changes all the lockers that are multiples of their numbers, that is, each locker is changed by each athlete who has a number that is a factor of the locker's number.
- All lockers visited an odd number of times, that is, locker numbers that are perfect squares because they have an odd number of factors, will remain open in the end.
- Any non-perfect square number will have an even number of factors, hence that locker number will remain closed.
- The locker numbers with the most factors would be changed the most times.
- All prime number lockers are switched only twice.

Students, of course, should be required to use properties of natural numbers to justify the results above.

 Activity 10 Exponential growth and decay

Once students have determined the general pattern behind exponential growth and decay, it is a good idea for them to see what it looks like graphically. They will be better able to explain what each parameter and variable represents in the general case $y = ca^x$ if they have looked at the graphs of multiple examples.

TEACHING IDEA 7: Demonstrating exponential growth and decay

This group activity will demonstrate the process of exponential growth. You will need 50 coins and one cup per group. Two coins represent stage 1. Place the two coins in the cup and toss. Count the heads that are showing. Add that many more coins to the cup. Record, in a table, the total number of coins in the cup after each toss. Repeat the process until all the coins are in the cup. Use a GDC or software to record the data on a scatterplot. Ask students to suggest real-life situations that behave in this way.

Repeat the activity to demonstrate the process of exponential decay. Start with the 50 coins in the cup (this is stage 1). This time, remove every head that is showing when the coins are tossed. Count the remaining coins at the end of each toss. Repeat the process until you have one coin remaining. Use a GDC or software to plot the data on a scatterplot. Ask students to suggest real-life situations that behave in this way.

After students have completed the activities in the student book, you could revisit these scatterplots. Ask students to use a GDC to determine the equation of the exponential model that describes the data.

An amusing variation of this activity is to use chocolates or sweets instead of coins, with one side of the chocolate or candy representing the "head". The students can eat the sweets during the decay activity as they are removed!

TIP

You have only considered the characteristics of the function when both a and c are positive numbers, as that is all that is needed given the context of the original questions. You could also do an inquiry into other transformations including negative values and inverses.

TEACHING IDEA 8: Choosing different values

INQUIRY Investigate how c and a affect the graph of $y = ca^x$.

Give the students these instructions.

a) Choose different values for $a > 1$. Write a general statement about the shape of the graph $y = a^x$ and the effect on the graph as the value of a increases, when $a > 1$.

b) Now choose different values for a that are greater than 0 but less than 1. Write a general statement about the shape of the graph $y = a^x$ and the effect on the graph as the value of a increases when $0 < a < 1$.

c) What is special about the y-intercept in all of these examples? Use your knowledge of the laws of exponents to explain this special characteristic.

d) Now add the parameter c to the function. What happens to the graph of $y = ca^x$ as the value of c changes? Choose some different positive values for c and then write a general statement about the effect of changes in the value of c on the graph $y = ca^x$ when $c > 0$.

Create a summary of the characteristics of the graph $y = ca^x$ given changes in the values of a and c.

Include a discussion of:

- equation of asymptotes
- domain
- behaviour of the graph as $x \to -\infty$
- behaviour of the graph as $x \to \infty$
- coordinates of the y-intercept
- range.

EXTENSION

Using logarithms to find generation times

The generation time is the time it takes for the cells to double in number (they divide into two) and is represented by the basic formula $G = \dfrac{t}{x}$, where G represents the generation time, t represents time (in minutes or sometimes hours, given the context) and x represents the number of generations.

Task: Ask the students to calculate the generation time of *E. coli* if it took 2 hours for the bacteria population to grow from 1000 to 60 000 cells in a laboratory test.

You can simply tell students to use the formula they derived for the doubling of bacteria, in combination with the basic formula above, and let them work independently. Alternatively, scaffold the task for them a little more. Hints such as: "Look for the variable that is common to both equations and substitute one into the other," and "Transpose the equation to make x the subject, using laws of logarithms," would guide students in the right direction.

Since they are dealing with the doubling of bacteria, the exponential growth formula is:

$$y = c2^x$$

c represents the number of bacteria at the beginning of the time period

y represents the number of bacteria at the end of the time period

x represents the number of generations (number of times the bacteria population doubles during the time period).

$\log y = \log c + x \log 2$

$x = \dfrac{\log y - \log c}{\log 2}$

$x = (\log 2)^{-1} (\log y - \log c)$

$x = 3.32 \log \left(\dfrac{y}{c} \right)$

Then substitute into the formula $G = \dfrac{t}{x}$.

So, the formula for generation time, given the numbers of bacteria and growth time, is:

$$G = \dfrac{t}{3.32 \log \left(\dfrac{y}{c} \right)}$$

Therefore, the generation time for *E. coli* in a laboratory is 20.3 minutes.

a) What are the differences between exponential patterns and polynomial patterns?

b) When determining what pattern is being modelled, how do you know it is represented by an exponential function and not a polynomial function?

TIP

This could be either an introductory activity for a unit on exponential growth and decay or a summative task at the end. It could also form a part of an interdisciplinary unit if students were asked to read and analyse *The Andromeda Strain* in either a language or sciences class.

TEACHING IDEA 9: Exponential growth in literature

In 1969, Michael Crichton, who would later go on to write *Jurassic Park*, wrote *The Andromeda Strain*, a chilling novel about an extraterrestrial micro-organism that threatens the human race. In his book, Crichton writes:

The mathematics of uncontrolled growth are frightening. A single cell of the bacterium E. coli would, under ideal circumstances, divide every twenty minutes. That is not particularly disturbing until you think about it, but the fact is that bacteria multiply geometrically: one becomes two, two become four, four become eight, and so on. In this way it can be shown that in a single day, one cell of E. coli could produce a super-colony equal in size and weight to the entire planet Earth.

Students could be given this quote and asked to prove whether Crichton was exaggerating. They would need to do some research on the surface area and weight of both the Earth and an *E. coli* bacterium and then model the situation with a graph and/or an equation. Students could also be asked to find exactly when the two would be equal, requiring them to solve exponential equations.

Activity 11 The Titius–Bode law

In this activity, students discover a pattern linking the distances of the planets from the Sun. Interestingly, planets orbiting other stars also seem to follow this same pattern.

Bode's law is an interesting pattern, since it is derived from a pattern of numbers.

Assessment

This activity is appropriate to use as a summative task. If you choose to assess students on this task, you can use criterion D. The task-specific descriptor in the top band (7–8) for Criterion D should read that the student is able to:

- write a recursive and explicit formula correctly for all arithmetic and geometric sequences
- write a general formula correctly for any sequences that are neither arithmetic nor geometric
- generate the next three terms of the sequence in the Titius–Bode law correctly
- justify the degree of accuracy of the three terms they generate
- justify whether the next three terms of the sequence makes sense
- justify whether they think Pluto should still be considered a planet.

Stage 1 of the unit planner

Key concept	Related concepts	Global context
Relationships	Patterns Quantity	Scientific and technical innovation
Statement of inquiry		
Relationships between quantities have helped us understand and predict patterns in the universe.		

 CHAPTER LINKS

Students can apply the formulae derived in Chapter 5 on Change, in the student book, here.

Describing the pattern with a formula is not as easy as it seems since students will eventually need to introduce a new parameter. The original sequence, 0, 3, 6, 12, 24, … is a geometric sequence. Students should be able to find an explicit and a recursive formula that generates terms after the first one $(n > 1)$. The next two sequences are neither arithmetic nor geometric but can be generated in a variety of ways.

In question **e)**, when students are asked to decide whether Pluto should have been demoted, many may notice that, in fact, Pluto is close to the position predicted by Bode's law and that it is Neptune that does not follow the pattern. Students can also be asked to research how the law was used to locate or discover celestial objects.

Some of the newer and more spectacular areas of mathematics in recent times are fractal geometry and chaos theory. While dynamical systems are not new, the mathematics to describe and understand these systems is new.

The purpose of the chaos game is to see how patterns can result from seemingly random events.

 Activity 12   **The chaos game**

This activity works well if students are paired off within the group. Give each group the same template to work on, preferably on either an overhead transparency or transparent paper. Alternatively, you could make this an interactive activity by visiting one of the websites mentioned in the web links box.

The template should have on it a large equilateral triangle, with vertices labelled L (left), R (right), and T (top).

It is important to go through the instructions in the student book with the students. Let them mark their diagrams for the first few moves, to make sure that they become familiar with the rules of the game.

When all the groups have made approximately 25 moves, collect all the transparencies and superimpose them on each other. If there is not enough evidence to begin to see a pattern, let the students complete more moves. Eventually, they should begin to see the Sierpinski triangle. This is a famous fractal, as is the Koch snowflake in Chapter 15 on Space.

WEB LINKS
Go to www.shodor.org (follow links to **Activities & Lessons>Interactivate> activities>The Chaos Game**) or http://school.discoveryeducation.com (search for The Chaos Game).

EXTENSION

a) Play the game with more vertices, guessing beforehand what figures might appear! Research famous fractals and their properties.

b) Encourage students to investigate the link between the Sierpinski triangle and Pascal's triangle.

Summary

Through the activities in this chapter, students have been introduced to some famous mathematical patterns in nature and in everyday objects. They have also used pattern generation to attempt to solve some algebraic and real-life problems. Recognizing inherent patterns is one of the first steps in solving problems. It is hoped that this chapter will serve to reinforce this very important problem-solving skill.

CHAPTER
12 Quantity
An amount or number

	ATL skills	Mathematics skills
TOPIC 1 Volumes, areas and perimeters		
Activity 1 Maximizing area	✓ Propose and evaluate a variety of solutions.	**Geometry and trigonometry** ✓ Calculate the dimensions of shapes, given some conditions, and determine values that maximize them.
Activity 2 Optimal design for conical paper drinking cups	✓ Create novel solutions to authentic problems.	**Algebra** ✓ Determine and manipulate expressions of one variable in terms of other specific variables given, with and without a GDC. **Geometry and trigonometry** ✓ Calculate the dimensions of shapes, given some conditions, and determine values that maximize them.
TOPIC 2 Trigonometry		
Activity 3 Angle investigation	✓ Make inferences and draw conclusions.	**Geometry and trigonometry** ✓ Use degrees and radians to express sizes of angles and recognize the advantage of the use of each unit.
Activity 4 Periodic motion and the sine function	✓ Make unexpected or unusual connections between objects and/or ideas.	**Geometry and trigonometry** ✓ Graph sinusoidal functions and analyse the effects of parameters of these functions on their graphs.
Activity 5 Transforming the sine function	✓ Select and use technology effectively and productively.	**Geometry and trigonometry** ✓ Graph sinusoidal functions and analyse the effects of parameters of these functions on their graphs.
Activity 6 Changing a tyre	✓ Make unexpected or unusual connections between objects and/or ideas.	**Geometry and trigonometry** ✓ Reflect on the meaning of parameters in a real-world context.
TOPIC 3 Age problems		
Activity 7 Mean, mode and median age	✓ Develop contrary or opposing arguments.	**Statistics and probability** ✓ Calculate mean, mode and median of student ages.
Activity 8 Histograms	✓ Understand the impact of media representations and modes of presentation.	**Statistics and probability** ✓ Conduct a graphical analysis of histograms. ✓ Recognize differences between continuous and discrete ungrouped and grouped data; in particular, analyse the impact of treating age as a discrete or continuous variable.
OTHER RELATED CONCEPTS	**Measurement** **Representation** **Space**	

Introducing quantity

In this chapter, students will start with a real-life problem that allows them to deal with familiar quantities: length, area, volume and angle size. Next, they will be given a task in which some quantities are supplied but others are unknown. As they begin to solve the problem they will need to make decisions about the level of accuracy required and the choice of scales when it comes to graphing relations between quantities. The move from a concrete situation, involving the use of materials that they can manipulate and measure, to defining an algebraic relation between variables, invites students to explore the situation in the abstract and transfer their knowledge to other contexts.

In Topic 2, students are challenged to work with a different unit to measure angles—the radian. They reflect on its advantages and consider the reasons why it is the unit of choice of mathematicians as well as the SI unit used by scientists. Students also explore the effects of changing parameters in trigonometric functions and how these same effects can be seen and analysed in cyclic phenomena such as the motion of a bicycle tyre.

Finally, in Topic 3 students revise the statistical treatment of discrete and continuous data and reflect on the implications of treating age as a discrete or a continuous quantity.

> *To criticize mathematics for its abstraction is to miss the point entirely. Abstraction is what makes mathematics work. If you concentrate too closely on too limited an application of a mathematical idea, you rob the mathematician of his most important tools: analogy, generality, and simplicity. Mathematics is the ultimate in technology transfer.*
>
> Ian Stewart

TOPIC 1 — Volumes, areas and perimeters

Understanding the relationships between quantities and expressing these as formulas are the basis for so much in mathematics. The ability to manipulate algebraic expressions and transpose algebraic equations is an essential skill when solving complex geometry problems, such as those found in Activity 2.

Activity 1 — Maximizing area

Whether students are looking at maximizing area given a set perimeter, or minimizing perimeter given a set area, the most important step is setting up the problem so that they are testing every possible combination of quantities. The best way for students to start is by using a spreadsheet and setting up the appropriate formulas. This can then be applied to more complex questions involving three dimensions and difficult formulas.

The first column can be the independent variable (in this case length). All the other columns will be connected to this variable through formulas. Students can then fill in, down the columns, to test for all possible combinations of length and width.

Length	Width	Area
This value will range from 0 to 30	$=\dfrac{60-2l}{2}$	$=l \times w$

Ask students to use the results from the table to draw graphs, so that they see how the quantities are represented visually. They can also identify the maximum point.

TEACHING IDEA 1: Moving to the algebraic form

To start a discussion about this task, ask students to consider the degree of accuracy that is necessary, for the lengths selected. Make sure they justify the number of decimal places they used for the length and the quantity by which they increased the length for each increment. Then introduce the idea of using algebra instead of numerical values for a dimension. In this way, they will cover every single possible quantity for each dimension.

Then set this task.

a) Call the length x. Find an expression to represent the width and the area of the rectangle

b) Use a GDC to graph the area of the rectangle in terms of x. Calculate the value of x that maximizes this area.

TIP

Activity 1 is quick and easy, and you can do it with the students. It will familiarize them with using spreadsheets to set up a systematic method for solving this type of problem. They can build on this idea to derive expressions for all dimensions in terms of one given variable. This should be helpful when students tackle a more complex problem (Activity 2), as they will have the confidence to set them up and work through the problem.

EXTENSION   **Why will a circle yield the greatest area?**

In this activity, students will determine that the square has the largest area, among all rectangles with the same perimeter. This is a useful starting point from which to determine what 2D shape in general will yield the largest area, given a fixed perimeter. Let the students determine what shapes to test. Ask them to justify their selections in terms of number of sides and regularity of side lengths. When they discover that a circle will yield the greatest area, given a fixed perimeter, ask them to justify why this is the case.

👤 Activity 2   Optimal design for conical paper drinking cups

In this task, students apply their knowledge of algebra and geometry to a manufacturing issue. They are asked to design the best possible template to create conical drinking cups.

Assessment

If you choose to assess students on this task, you can use criterion D.
The task-specific descriptor in the top band (7–8) should read that the student is able to:

- set up appropriate charts (spreadsheets) to solve the problem
- find the correct expressions to represent the radius of the base and height of the cone in terms of the angle of the cut-out sector
- derive the correct general algebraic volume formula of the cone in terms of the angle of the cut-out sector
- graph the scenario correctly
- determine and explain all key points on the graph
- use a GDC or graphing software to correctly determine the exact values to maximize volume
- discuss the restrictions on the domain, given the context of the problem
- evaluate alternative scenarios of cone size
- justify the degree of accuracy that would be needed in the context of the problem
- reflect critically on the template, given the context of the problem and its suitability in real life and justify their recommendation.

Key concept	Related concepts	Global context
Relationships	Quantity Measurement	Scientific and technical innovation
Statement of inquiry		

Understanding the relationship between quantities improves technical innovation in the development of products.

STEP 1 **Exploring different templates**

Students will try to determine the relationship between the value of the angle of the sector removed and the volume of the cone. Hence, they will determine the size of sector that will give the greatest volume.

To complete step 1, students must know the volume formula for a cone. The formula is easily derived through a quick inquiry into how many cones will fit into a cylinder with the same dimensions. It is also a good refresher of the formula. The most effective way to do this is to fill the cones with water and simply pour this water into the cylinder.

STEP 2 **Algebraic and graphical representation of the volume**

Students are asked to find an expression to represent the volume of the cone in terms of the size of the cut-out sector. Then they have to calculate the size of the cut-out sector that maximizes the volume.

Students must be able to transpose and simplify formulas for this task. Before they start, give them plenty of practice in substituting expressions into formulas and rearranging formulas to make different variables the subject.

REFLECTION

Whenever students complete mathematical solutions to authentic problems, they should reflect on what they have done, considering its feasibility.

This task has a high level of difficulty. Perhaps, before students begin the activity, you could work through a similar problem with the class, taking the opportunity to highlight and emphasize key elements. Problems in which students investigate, for example, how to fold a rectangular sheet of plastic measuring 50 cm by 20 cm into a rectangular prism, so that it holds a maximum volume, are useful as preparation for this activity.

TIP

When writing the report, students can take a screenshot of the GDC screen to help them explain their solution. It also makes it easier for them to move from the algebraic to the graphical representation. An emulator of the GDC can be downloaded free onto any computer to make it easy to copy and paste these pictures. Alternatively, they can connect the GDC to their computer via a cable.

Why use radians to measure angles?

In the student book, there is some information about mathematicians who first used radians, and reference to the use of sine tables to solve problems before the calculator era. As students often wonder about how trigonometric calculations were performed before the invention of calculators, it could be an interesting group activity to discover more about this.

TIP

You can challenge students to build a protractor that measures the angles in radians. They should decide about the number of divisions needed to allow accurate reading of the angle sizes.

TEACHING IDEA 2: Radians and degrees

INQUIRY An inquiry approach to the activities in this topic may make them more interesting. Students can easily do a quick research activity to find out about the SI units in general and about the radian as the unit to measure angles in particular.

Although, while researching, students may quickly find that $1 \text{ rad} = \dfrac{180°}{\pi} \approx 57.2958°$ it may help them to understand this relation better if they actually deduce it and explain it to their peers. They can then be asked to apply the relation and convert quantities expressed in degrees to radians and vice versa.

This class activity can be done in groups.

Instructions

a) Research radians and degrees and write down a clear explanation for the formula:

$$1 \text{ rad} = \frac{180°}{\pi} \approx 57.2958°.$$

b) Deduce that $1° = 0.01745\ldots$ rad.

c) Use the conversion relations above to convert these quantities into the required units.

(i) $35° = $ _____ rad   (ii) $4 \text{ rad} = $ _____°   (iii) $3.6 \text{ rad} = $ _____°

d) Explain your decision about rounding in part **c)**.

To explore the issue of accuracy when working in radians further, it is worth spending time using protractors to measure a few angles in degrees and converting the quantities into radians.

Activity 3   Angle investigation

One aspect of this activity is that students should begin to think about and refine the concept of angle—from angle as a surface to angle as a number.

To complete this activity, students will need protractors of different sizes. A good option is to photocopy a protractor on overhead transparencies, using different zoom options (enlarge or reduce it by a different factor each time).

As the students follow the instructions of the activity they will quickly notice that $\dfrac{l}{r} = \alpha$ which leads to the conclusion that the size of an angle, when measured in radians, is given by a ratio of two lengths and is therefore a number, that is, an abstract quantity that is dimensionless!

The concept of an angle being a number is abstract but very important. If students are to be able to decide when they need to use radians, it may be worth exploring this concept further and even challenging students' concept of an angle.

After all, what is an angle? Is it a surface? A length? Or, a number?

Ask students what they think an angle is, or ask them to "draw" an angle (that is, a representation of an angle). You may get different answers and examples.

DP LINKS

In the Diploma Programme, both SL and HL mathematics courses include study of trigonometric calculus, which requires the use of radians.

- If you think of internal angles of a triangle you are thinking of a portion of the plane.
- If you think of the movement of a point on a circle you are thinking of a length.
- The activity in the student book leads students to conclude that an angle is a number!

When it comes to drawing an angle there is no rule about the length of the sides. Usually, you simply draw a V-shape of any size to represent an angle.

Ultimately, an angle is an abstract concept with multiple real-life applications, as highlighted in future activities.

Activity 4 — Periodic motion and the sine function

This activity is a novel way of introducing the sinusoidal graph, since it is produced by something that is probably familiar to most students. They are asked to measure the height of the valve of a bicycle tyre as the wheel is rotated and then stopped at regular (equal) intervals. This will produce a natural sinusoidal curve that students can analyse.

While the height of the valve is clearly the dependent variable, students often wonder what to use as the independent variable. One of the most common suggestions is to use the spokes to measure the rotation. The height of the valve can be measured after the wheel has turned and a set number of spokes have passed a fixed point. Alternatively, students can push the bike forward and stop after a set short distance has been covered, at which point they will measure the height of the valve. They will then simply repeat the process until the wheel has moved through one complete rotation.

When students have produced a sinusoidal curve, they can research the meanings of the terms "period", "amplitude" and "line of equilibrium" or "midline". They can use what they have found out to define the properties of the curve.

You could ask students to begin measuring with the valve at different points. For example, some may be asked to start with the valve at its lowest point, some at its highest and others with the valve at the same height as the axis of rotation (the axle). They can then be asked to determine what, if any, effects this has on the amplitude, period and line of equilibrium.

 **WEB LINKS**
Enter "Stefan Waner sinusoid" into a search engine to see an animation of a sinusoid produced by a rotating bicycle tyre.

When students graph the sine function in part **e)**, it is important that they understand how the angles in the table relate to one another, so that they can draw a proper scale on the horizontal axis. Suggest that they convert the angles into degrees to show that they are, in fact, equally spaced already. If students cannot see that the graph will repeat itself, let them add more rows to the table and continue finding the sine ratio on their calculators and graphing these new points. Note that while this approach leads students to use a calculator to determine the value of the sine function, they can also use exact values if they have the prerequisite knowledge.

Activity 5 — Transforming the sine function

After discovering the sine function and the properties of its graph, students are ready to think about transforming the graph through the use of parameters. In this inquiry-based activity, they explore how changing individual parameters in the sine function affects its graph. They are led first to make a conjecture and then use a GDC to test it. It is important that they understand the nature of the changes individually before moving on to the exercises, where they will then have to analyse multiple changes at once. Students who are experiencing difficulty may have one of the following issues:

- their calculator is in the wrong mode (degree versus radian)
- the "window" for the GDC does not allow students to see all of the values of the dependent and independent variables.

TEACHING IDEA 3: The cosine function

INQUIRY Let students perform the same tasks but with the cosine function, starting with part **e)** in Activity 4 and continuing through Activity 5. Students can then be asked to compare and contrast the graphs of sine and cosine functions.

DP LINKS
Trigonometric functions are part of all mathematics courses in the DP but they are studied in greater detail at HL.

EXTENSION

While the horizontal shift parameter is not explicitly studied here, students can perform similar experiments to see its effects. This could lead to a discussion on how the sine and cosine graphs are simple horizontal transformations of each other.

Activity 6 Changing a tyre

Having seen that the motion of a bicycle tyre can be represented by a sinusoid, students are ready to determine how modifying the tyre will affect the graph. Students follow up their work in Activity 4 by drawing the graph of $y = \sin x$ (with x measured in radians) and they can then make the link between the motion of a tyre and the sine function. First, they will find the equation of their sinusoid and then analyse how making modifications to the bicycle tyre itself can change the amplitude, period and line of equilibrium.

Let students carry out research on the features of different models of bike, such as BMX, mountain bike and road bike tyres.

- What quantities are used to describe the tyres?
- What features are designed for speed?
- Which ones provide more stability?

∞ INTERDISCIPLINARY LINKS

Trigonometric functions have applications to a wide variety of phenomena. Some students may also find the study of tides, the amount of daylight over the course of a year, the phases of the Moon, the population of arctic lynx and electricity to be equally engaging. Problems, test questions and summative tasks can be designed from any of these applications.

TOPIC 3 Age problems

Collecting and analysing data, both numeric and categorical, is fundamental to statistics. Numeric data can be further classified as either "discrete" or "continuous", an important step, as the type of data can influence the type of analysis that can be performed. However, for such a seemingly simple classification system, it is not always so straightforward.

 Activity 7 Mean, mode and median age

In this activity, students will explore age and discuss whether it can be seen as discrete or continuous as there is no clear answer to this.

TEACHING IDEA 4: Discrete or continuous data

INQUIRY Before even beginning the activities in Topic 3, present students with examples of continuous data and examples of discrete data. Ask them to make conjectures about their definitions and then test these definitions with other examples. (This can be done as a whole class or in groups with envelopes of examples given to each group.) Students continue to modify their definitions until they correctly identify every new example. At this point, ask them to add two more items to each list. This same type of activity can be done to help students discern the difference between numeric and categorical data, although they often recognize the difference just from the names.

TIP

Students should be able to explain that continuous data must be measured while discrete data must be counted.

When students understand the difference between discrete and continuous data, ask them to find examples in which continuous data is grouped or represented as categorical data (for example, "teenagers", "senior citizens" and so on). Use these discussion questions.

- Why would this be done?
- What is gained or lost when this is done?

The study of data and statistics often generates very useful conversations and debates in class. Showing students examples of mathematics in everyday life is an effective way to engage them and give context to their learning. Ask them to classify different data, such as:

- scores in a video game
- shoe sizes
- prices of food
- television channels
- grades in school
- number of text messages sent
- length of a Skype call
- your place in a race (first, second, and so on)
- votes in an election.

 Activity 8 **Histograms**

In this activity, students consider how data is presented, given that it may be continuous or discrete. Based on their answers to parts **d)** and **e)**, students could be asked to represent the data in a way they think makes the most sense and then share it with the class.

Summary

Students have explored different types of quantity: those that can be expressed as a magnitude, number or variable. While working with many different units, students have realized that all quantities are measurable and that they need to make decisions about level of accuracy when they are working through problems. They must also take care with their choice of scales when it comes to graphing relations between quantities. They have discovered that a good understanding of how to calculate and represent quantities will lead to efficient problem-solving, and accurate and appropriate solutions.

Representation

The manner in which something is presented

	ATL skills	Mathematics skills
TOPIC 1 Points, lines and parabolas		
Activity 1 A parabolic times table	✓ Consider ideas from multiple perspectives.	**Number** ✓ Use points on the parabola and the *y*-axis to generate the times table. **Algebra** ✓ Graph parabolas. ✓ Represent points and lines graphically.
Activity 2 Graphical representation of lines	✓ Consider ideas from multiple perspectives.	**Algebra** ✓ Use set-builder notation.
Activity 3 Paths of ships	✓ Consider ideas from multiple perspectives.	**Algebra** ✓ Solve problems involving position, velocity and time and intersection of paths of moving objects when represented by Cartesian and vector equations of lines.
TOPIC 2 Probability trees		
Activity 4 Probability trees game	✓ Revise understanding based on new information and evidence.	**Statistics and probability** ✓ Use probability trees to represent a situation and calculate the probability of selected outcomes.
Activity 5 Conditional probability and medical test results	✓ Evaluate and manage risk.	**Statistics and probability** ✓ Use probability trees to represent a situation and calculate the probability of selected outcomes. ✓ Use probability trees to calculate the conditional probability of a real-life situation.
TOPIC 3 Misrepresentation		
Activity 6 Standard deviation	✓ Understand the impact of media representations and modes of presentation.	**Statistics and probability** ✓ Use and interpret measures of central tendency and measures of dispersion. ✓ Interpret graphs.
Activity 7 Measures of central tendency	✓ Understand the impact of media representations and modes of presentation.	**Statistics and probability** ✓ Use and interpret measures of central tendency and measures of dispersion.
OTHER RELATED CONCEPTS	Quantity Simplification	

Introducing representation

Every student needs to be able to move between different forms of representation as well select the most appropriate one in a given context.

The algebraic representation of a function may not be the best form to work with when students are trying to understand its characteristics. Encourage students to graph functions and then look at key points on their graphs to see how they translate into algebra. From this, they will start to understand functions and make generalizations about their specific properties.

> *Beauty is the first test: there is no permanent place in this world for ugly mathematics.*
>
> GH Hardy

TOPIC 1 Points, lines and parabolas

This topic is intended to give students a deeper understanding of the various ways of representing straight lines. It should raise their awareness of the limitations of Cartesian equations versus vector equations of lines as they discover the advantages of selecting an appropriate form of representation of lines to solve specific problems efficiently.

You could use the activity below to help students understand the advantages of being able to use and move effectively between different forms of representation. Asking them to write a commentary gives them practice in communicating their understanding of how to interpret a graph. They will investigate key points of a linear, a quadratic and a cubic function in multiple representations, in an amusing context.

TEACHING IDEA 1: The amazing race

Suppose you are a commentator making a podcast to describe an animal sporting event. The functions below represent three of the contestants in a 200 m race.

$$d_1 = 11t$$
$$d_2 = \frac{1}{2}t^2 + 20t$$
$$d_3 = (t-2)(t-9)(t-18) + 20$$

where d_1, d_2 and d_3 are the distances, in metres, travelled after t minutes of the race.

Look at the points of intersection, turning points, x and y intercepts and slopes to help you to prepare an accurate commentary of the race and decide which animals are involved. You should explain:

- the characteristics of each curve in detail
- which animal won the race
- the placement of the other two animals and their respective finishing times.

You can use technology to create an appropriate graphical representation of the shapes of these functions to help you explain your story.

In your commentary, you must include:

- a classification of each type of mathematical equation
- a determination of which animal represents each equation, and why
- a clearly labelled graph that shows these three functions on the same set of axes
- a full set of mathematical solutions that you used to determine which animal won the race and all the finishing times
- a discussion of how different forms of representation were used to help complete this task—when it was better to use graphical forms and when using algebra was more appropriate.

The *Mathematikum* is a fascinating museum in Germany. One of the exhibits is the parabolic times table activity (the first activity in this chapter of the student book). The purpose of the exhibit is two-fold: to demonstrate how to use a parabola in multiplying two numbers graphically, and to provide a pictorial representation of the famous sieve of Eratosthenes for identifying prime numbers.

The activity is quite straightforward. The first four questions in Step 1 should be accessible to all students. In question **e)**, students are asked to explain why the graph works as a "times table". If necessary, give them a hint. Suggest that they consider similar triangles, as demonstrated in the diagram below.

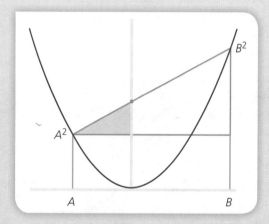

To form an equation, they should be able to use the fact that the shaded triangle and the red triangle are similar.

For Step 2, students will obtain a picture similar to the one in this diagram. They can easily see that the numbers on the y-axis that are not y-intercepts of the lines are indeed prime numbers.

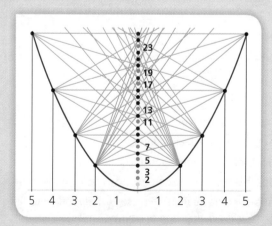

Activity 2 Graphical representation of lines

This activity is designed to encourage students to reflect on the meaning of a Cartesian equation of a line. What is the meaning of the equation $x + y = 4$? By writing on each point on the grid, the value of $x + y$, students can quickly spot the patterns formed by all 4s.

Students can do this activity individually, on paper, or as a class activity, on the board or using software (DGS). The activity can be a good starting point for discussions.

- This method is simple, but is it an efficient method?
- What are its limitations?
- Is it easy, if possible at all, to use it to find intersections of lines?

The activity also offers a good opportunity to introduce or revise set-builder notation, as students look at lines as sets of ordered pairs of real numbers. This can promote clear and good communication through the use of correct mathematical notation.

The analysis of an expression such as $A = \{(x, y) \mid x + y = 4\}$ allows clarification of many aspects of mathematical language.

- A is a set of points represented by a capital letter.
- (x, y) is an ordered pair of real numbers. When a domain is not specified, students may assume that it is the set of real numbers.
- The vertical line symbol means "such that" and you use it when you want to impose a restriction on the values of the variables.
- $x + y = 4$ in this case restricts the collection of points to those with coordinates that satisfy the condition—this equation defines the line, in that the points on the line are exactly the points on the plane with coordinates that are a solution of the equation.

It is also important to highlight that the line is a continuous line, even if, when you find some of its points, you tend to choose those with integral coordinates. It may be worth offering students the opportunity to plot other types of set with similar algebraic representations.

TEACHING IDEA 2: Drawing representations
Set students this task.

Draw the geometrical representation of each of these sets on a Cartesian plane. Describe each set, in words.

a) $\{(x, y) \mid x + y = 4 \text{ and } 0 \leqslant x \leqslant 4\}$ (a line segment)
b) $\{(x, y) \mid x + y = 4 \text{ and } x > 0\}$ (a ray)
c) $\{(x, y) \in \mathbb{N} \times \mathbb{N} \mid x + y = 4\}$ (a discrete set of points)
d) $\{(x, y) \in \mathbb{Q} \times \mathbb{Q} \mid x + y = 4\}$ (a discontinuous but dense set of points)

This activity can be extended and used as an informal activity. Students can work in groups and challenge each other to describe in words a set, using set builder notation.

TEACHING IDEA 3: A game to be played in pairs
Each player draws two Cartesian grids with both axes numbered from 0 to 10. Each player draws a rectangle of any size on one grid. The vertices must have integral coordinates. Neither player shows this rectangle to the other player.

Players take turns to try to determine or guess the position of their opponent's rectangle, by proposing conditions expressed in set-builder notation. The other player can just answer: "This set intersects the rectangle," or "This set does not intersect the rectangle."

The goal is to discover the position of the opponent's rectangle.

Step 3 in particular allows students to use different Cartesian equations for the same line.

- Equations of the form $ax + by = c$ are particularly useful to determine axes intercepts (the x-intercept $= \dfrac{c}{a}$ and the y-intercept $= \dfrac{c}{b}$ for $a, b \neq o$)
- Equations in gradient-intercept form $y = mx + c$ show explicitly the value of the slope (gradient) m and the y-intercept c.

TIP
When two points are given, as in Step 3, the regression line provides an exact fit as a line is defined by exactly two points. This allows students to use the GDC to check their answers when working independently. In Step 5 students can explore this idea further and look at a situation where three non-collinear points are given.

Activity 2 includes a real-life and useful application: conversion between temperature scales used in different regions. This is a good opportunity to talk about the use of different units in different parts of the world. Currency conversion problems can also be used to explore applications of linear relations and a variety of conditions can be introduced (fixed conversion fee or percentage fees). This gives students an opportunity to share information with the class about their own home countries and to promote intercultural awareness. Alternatively, students can research online about the currency of other countries they are interested in, and conversion costs, and use what they learn to estimate the cost of holidays in these places.

TEACHING IDEA 4: The role of parameters

The role of the parameters in equations is better explored with DGS or with GDC, using sliders or with free applets available online. These tools provide in real time many examples to analyse, as parameters are animated and allow students to discover the effects of their change on the graphs of the lines.

In step 5 they are challenged to experiment with different strategies to discover the parabola defined by these points. Graphs of quadratic functions (parabolas) are explored in detail in Topic 1.

⊂⊃ **WEB LINKS**

For instructions on using sliders with the TI-Nspire, see www.dummies. com and search for "how to work with sliders on the TI-Nspire". Alternatively, you might want to use software such as Geogebra or Autograph.

EXTENSION

Challenge students to investigate the type of curve defined when two, three, four, … points are given. They should use GDC or DGS and regression analysis. Then guide them to relate the number of points with the number of coefficients to be determined to define the equation of the curve:

- two points define a straight line with an equation of the form $y = a_1 x + a_2$ (for non-vertical lines)
- three (non-collinear) points define a parabola with an equation of the form $y = a_1 x^2 + a_2 x + a_3$ and so on.
  They should find that the number of points matches the number of coefficients to be determined.

👤 Activity 3   Paths of ships

This activity focuses on vector equations of lines versus Cartesian equations of lines, and on the importance of parameters when lines are used to model situations that involve a change of conditions or movement. It is designed to produce a simulation of a simplified real-life situation. In this case, the simplification results from the assumptions that the ships move along a straight line and that their velocity is constant (both in direction and magnitude). You may want to motivate students by talking about computer games in which the movements of the objects are controlled by several parameters: their velocities may vary in direction and intensity and depend on multiple parameters that the player and/or the computer controls.

⊂⊃ **DP LINKS**

In the Diploma Programme, vector geometry is explored further and extended to 3D, allowing application to more complex real-life situations.

Encourage students to work through the example before doing the activity in the student book. After this, they should be able to complete the activity independently, provided that they are familiar with:

- the relation between Cartesian and vector equations of lines and how to move between forms
- calculating the magnitude of a vector
- the distance between points as the magnitude of the displacement vector
- simultaneous equations and their use to determine intersection of lines.

This activity offers the opportunity to explore further the correct use of mathematical notation as it involves working with scalars and vectors. Students from different backgrounds may be used to different notation for vectors (or different resources used by students may use different notation for vectors). This activity uses the notation for vectors that appears on Diploma examinations.

This activity also uses precise command terms and it is worth clarifying their meaning. For example, in question **c)**, students are asked to deduce an expression. This requires that a logical sequence of steps is shown, starting from the vector equation of the line and manipulating it to obtain the required Cartesian equation.

From $\begin{pmatrix} x \\ y \end{pmatrix} = \begin{pmatrix} 0 \\ 28 \end{pmatrix} + t \begin{pmatrix} 6 \\ -8 \end{pmatrix}$ you can obtain the two parametric equations of the line:

$\begin{cases} x = 0 + 6t \\ y = 28 - 8t \end{cases}$ (parametric because each variable x and y is expressed in terms of the parameter t).

Solving simultaneously and eliminating the parameter t, you obtain the Cartesian equation $y = 28 - 8 \times \dfrac{x}{6}$ that can be rearranged to obtain $4x + 3y = 84$.

> ## TEACHING IDEA 5: The interpretation of "define"
> The mathematical term "define" can at times confuse students when they are studying equations of lines and curves. On the one hand, "define" means "determine uniquely"—so each linear equation defines a line; on the other hand, the same line may be defined by many different equations (vector or Cartesian equations). To eliminate possible misconceptions, it is worth asking the questions in part **c)** in the following way.
>
> - How do you know that the equations $y = 28 - 8 \times \dfrac{x}{6}$ and $4x + 3y = 84$ represent the same line?
> - Which type of manipulations are you allowed to perform to obtain equivalent (linear) equations?
> - Can you say that the vector equation $\begin{pmatrix} x \\ y \end{pmatrix} = \begin{pmatrix} 0 \\ 28 \end{pmatrix} + t \begin{pmatrix} 6 \\ -8 \end{pmatrix}$ is equivalent to the Cartesian equation $4x + 3y = 84$ because they represent the same line?

As they work through this activity, it is important that students notice these points:

- When moving from a vector to a Cartesian equation of a line, they lose information, namely the parameter t is eliminated and therefore, in the context of the problem, the information about when the ship is at a particular point, is lost.
- For the reason above, the vector and Cartesian equations are not equivalent as they cannot even work backwards and obtain the vector equation from the Cartesian one.
- In context, it is also clear that two vector equations of the same line may represent different situations: for example, the equation $\begin{pmatrix} x \\ y \end{pmatrix} = \begin{pmatrix} 6 \\ 20 \end{pmatrix} + t \begin{pmatrix} 6 \\ -8 \end{pmatrix}$ is not equivalent to $\begin{pmatrix} x \\ y \end{pmatrix} = \begin{pmatrix} 0 \\ 28 \end{pmatrix} + t \begin{pmatrix} 6 \\ -8 \end{pmatrix}$ although they both represent the same line.

There are a number of possible extensions to this activity.

EXTENSION

Explore vector and Cartesian equations of perpendicular lines in the same context and add this extension to the activity.

a) Another ship, *Martinica*, departs from *ORT* at 15:00. *Martinica* follows a path perpendicular to *Cargy*'s path. *Martinica* travels at a constant speed of 12 km/h.

  (i) Find the velocity vector of *Martinica*.

  (ii) Write down the vector equation of its path, using the same parameter t.

  (iii) Show the path of this ship on the same graph that you used for your simulation of the situation.

b) Determine the position of each ship at 16:00.

You can also challenge students to explore further equations involving parameters. For example, using simulation, investigate the shape of the path of a point moving at non-constant velocity, such as $\begin{pmatrix} x \\ y \end{pmatrix} = \begin{pmatrix} 6 \\ 20 \end{pmatrix} + t \begin{pmatrix} 0.1 \\ -0.2 \end{pmatrix}$

REFLECTION

Topic 1 ends with a reflection in which students are first asked to summarize their learning about equations of lines, then to analyse the advantages and disadvantages of each form of representation and, finally, provide examples in a real-life context where a particular form of representation is more useful.

TOPIC 2   Probability trees

Probability trees are a very effective way of representing outcomes and their associated probabilities. In this topic students are introduced to conditional probability. In the first activity they look at the probabilities involved in a simple game. In the second activity, they apply conditional probabilities and tree diagrams to the analysis of a potentially controversial medical test.

Activity 4   Probability trees games

How can you tell if a game is fair?

Students can use probability trees to determine if a game is fair—if all players have an equal chance of winning.

Your students could play this game, as a practical exercise in probability.

Rock, paper, scissors is a game usually played by two people in which both players form one of three shapes (rock, paper or scissors) with their hands at the same time.

- Rock defeats scissors.
- Scissors defeats paper.
- Paper defeats rock.

If both players pick the same shape then the game is tied.

Predict if you think the game will be fair or not.

Construct a probability tree of all possible outcomes, to determine if this is a fair game.

Make generalizations about how to use the tree effectively to represent all of the probabilities. Base your ideas on your probability tree for all outcomes.

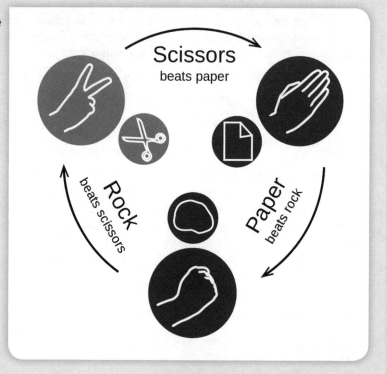

REFLECTION

- Can you think of other ways this information could be represented?
- Is a probability tree the best way to represent all possible outcomes and their probabilities?
- If you wanted to add a fourth shape, what rules would you have to alter and/or include to ensure the game is fair?
- Because of the "human element" of this game, it can be debated that you can have an advantage over your opponent by analysing their previous choices and making predictions based on patterns you can see in their selections. How could you alter this game to ensure the "human element" is removed but the basic rules and outcomes of the game still apply?

TEACHING IDEA 6: Games of chance

There are many games of chance that have been played throughout the centuries by different cultures in countries worldwide. Students could research some of these games and select one to analyse. Then they could construct a probability tree of all possible outcomes to determine if it is a fair game.

When analysing a game of chance, students could work individually or in pairs. You may choose to let them present and give a brief explanation of their game, and then determine if the game is fair based on the probability tree they constructed of all possible outcomes. You could also let students select games that they can recreate and play in class.

TEACHING IDEA 7: Creating a game

Ask your students to create their own game and submit a report including:

- a description of the game
- how to play the game
- all rules of the game
- all mathematical probabilities of the game represented in a probability tree
- an explanation of mathematical reasoning associated with the construction of the game (for example, the smallest section on a spinner would have the highest reward because the probability of landing on that sector is lower).

Activity 5 · Conditional probability and medical test results

Assessment

If you choose to assess students on this task, you can use criterion D.

The task-specific descriptor in the top band (7–8) should read that the student is able to:

- identify the relevant elements needed to determine the probabilities associated with the medical test
- set up an accurate probability tree to solve the problem
- apply appropriate mathematical strategies to correctly determine the probabilities associated with the test
- justify the degree of accuracy of the solution
- critically reflect on the probability results, given the context of the problem and justify their comments regarding its effectiveness.

Stage 1 of the unit planner

Key concept	Related concepts	Global context
Form	Representation Simplification	Identities and relationships
Statement of inquiry		

Representing data and statistics in the most effective form can help us better understand and evaluate health issues on a local and global scale.

Students will use probability trees to calculate the conditional probability of a real-life situation.

The easiest way to move from independent to dependent events is to investigate problems involving replacement, then problems involving no replacement.

Introduction activity

Assume that you have a bag of four marbles—red, blue, purple and green—and you select one then put it back in the bag, then select another. Construct a probability tree to represent the possible outcomes and probabilities of what marbles you will select.

Do this activity with replacement first, then move to no replacement as an introduction to conditional probability.

Highlight the differences between the two and show how the probabilities change.

Please be aware that this question deals with sensitive material. Take care when discussing the calculations and reflecting on their significance.

This task seems complicated but setting up the probability tree makes the mathematics easy to understand.

Students will need to research the approximate population and number of people diagnosed as HIV positive in their country in order to determine some of the probabilities on their tree—you can find those statistics beforehand to save time in the classroom if you wish.

WEB LINKS
You can do this with an actual bag of marbles, or comparative objects, or you can use the simulator at http://www.shodor.org and search for "interactive marbles".

DP LINKS
Conditional probability is covered in detail in the mathematics SL and HL courses, so an introduction to it here serves as a good foundation.

TOPIC 3 Misrepresentation

This topic has the potential for provoking interesting conversations and debates, as students analyse how data is presented and how it can then be interpreted. What is important here is that students begin to question data, both in terms of how it is collected and how it is reported and used.

The activities in this topic may be performed in any order. Students may be more familiar with the content in Activity 7, so you may prefer to start there. Regardless of the order in which the activities are tackled, it is a good idea to try to bring in as many real-life examples of how these statistical measures are used in the media, such as advertising and news articles.

Activity 6 Standard deviation

While students may be more familiar with measures of central tendency, which they will explore in Activity 7, many may know little about measures of dispersion. It would be a good idea to introduce some of the terms, such as range and standard deviation, and ask students to explain the difference between measuring dispersion and central tendency. They should also develop an understanding of what it means to be "within one (or two) standard deviation of the mean" in order to complete the activity.

TEACHING IDEA 8: Calculating σ

Show students how to calculate the standard deviation from raw data. Use a small set of data, in which the mean is an integer. Showing them how to calculate it gives them a clear understanding of what it means. They may prefer to see the mathematical formula after they have done the calculation several times.

Ask students to find the three measures of central tendency, as well as the range and standard deviation, for the data set 1, 4, 5, 6, 6, 7, 9, 10.

To simplify the calculation of the standard deviation, let them fill in a table like this one.

x	$x - \bar{x}$	$(x - \bar{x})^2$
1		
4		
5		
6		
6		
7		
9		
10		

TIP

Ask students why they need to square the last column. Why not just take the difference between the value and the mean value as is?

TEACHING IDEA 9: Standard deviation

When your students have worked on some standard deviation problems for themselves they could try these questions.

1. Suppose you need to register for a mathematics course, for which the passing grade is 70%. You have found this information about two possible courses that you could join.

   Class A: mean grade = 74%, $\sigma = 4$

   Class B: mean grade = 72%, $\sigma = 12$

   a) Describe the relative abilities of the students in each of those classes.

   b) Which class would you register for? Explain.

   The choice would be a very personal one and would depend on a variety of factors, such as whether or not students simply want a good chance of passing or whether they think they'd like to earn a higher grade.

2. These are test results out of 20 for two groups of students.

   Group A: 20, 20, 10, 0, 0

   Group B: 10, 10, 10, 10, 10

   a) Describe the relative abilities of the students in each of those groups.

   b) Suppose these two groups were in a competition. Which group do you think is more likely to win? Explain.

   Students may have different answers, depending on how the competition is set up and how many questions the groups are allowed to answer, but the students can clearly see that while the mean is the same, the measures of dispersion are significantly different.

When interpreting the graph, a case can be made for or against global warming. Because the level of ice is decreasing, something must be affecting it, which could be global warming. (It is important for students to be reminded that there is no way that data can show causality, that a change in one variable caused a change in the other.) However, it's also possible to interpret the graph as saying that, while the amount of ice is decreasing, it always seems to follow this pattern. The decrease doesn't necessarily have to be out of the ordinary and, in fact, the decrease could be due to other factors.

EXTENSION

Once students are comfortable with the notion of standard deviation, they could explore the standard deviation of a sample as opposed to a population, the normal distribution and z-scores.

TIP

Once the standard deviation is included on the graph, students should then be able to interpret the 2012 sea ice extent to be "out of the ordinary", since it is more than two standard deviations away from the mean. However, it is still not possible to say definitively that it is due to global warming because it could have been caused by other factors. Students could then be asked to research and interpret other data that would clarify their own position on global warming.

If students are well acquainted with the mean, median and mode of a set of data, they can begin this activity immediately. While students often find it easy to determine the measures of central tendency, they aren't always aware of when it is appropriate to use them.

TEACHING IDEA 10: Giving some real-life context

Newspaper and magazine articles that refer to the mean, median and mode can be brought in (by the students, teacher or librarian). Students can use these as an introductory task. They could be asked to report on when each measure was used and to explain why they think it was chosen over the other two. Students can then be directed to read the weblink in the student book and answer questions **b)**, **c)** and **d)**.

TEACHING IDEA 11: Two lies and a truth

Ask students to create three headlines for news stories based on data that they collect or find online. Each headline should be followed by a brief story that includes some "facts" based on a different measure of central tendency. In two of them they should use the most appropriate measure of central tendency while in the other they should not. Students then present their stories and try to convince their classmates. (For example, "Starting salaries on the rise!" *Based on the average salaries of 1000 men and women …*)

Statistics, especially descriptive statistics, is one of the branches of mathematics where students can find all sorts of interesting data and topics to debate. At the heart of these debates is often the notion of causality and whether that can actually be proven.

Summary

Understanding multiple ways of representing the same mathematical information and selecting the most appropriate form of representation are essential skills that support and improve students' ability to work through problems and communicate their ideas and solutions. By working through examples that included representing the real numbers and different representations of lines, constructing tree diagrams and critically analysing data and graphs, students have developed their ability not only to represent their mathematical ideas in an effective way but also to visualize those ideas as they worked through the problems.

Simplification

The process of reducing to a less complicated form

	ATL skills	Mathematics skills
TOPIC 1 Simplifying algebraic and numeric expressions		
Activity 1 The geometric mean	✓ Apply skills and knowledge in unfamiliar situations.	**Algebra** ✓ Identify terms in arithmetic and geometric sequences. ✓ Simplify using the laws of logarithms and exponents.
Activity 2 Dimensional analysis	✓ Make connections between subject groups and disciplines.	**Algebra** ✓ Simplify using the laws of exponents.
TOPIC 2 Simplifying through formulas		
Activity 3 Metric relations in a right-angled triangle	✓ Apply existing knowledge to generate new ideas, products or processes.	**Geometry and trigonometry** ✓ Calculate metric relations in right-angled triangles. ✓ Use the properties of similar triangles to find unknown measures.
Activity 4 Deriving formulas for other problems involving right-angled triangles	✓ Apply existing knowledge to generate new ideas, products or processes.	**Geometry and trigonometry** ✓ Use basic trigonometric ratios and/or the sine rule to derive formulas to find the heights of objects.
TOPIC 3 Simplifying a problem		
Activity 5 Polynomial equations	✓ Analyse complex concepts and projects into their constituent parts and synthesize them to create new understanding.	**Algebra** ✓ Factorize quadratic expressions and solve quadratic equations. ✓ Divide polynomials. ✓ Factorize higher degree polynomials and solve polynomial equations.
Activity 6 Dating the Earth	✓ Process data and report results.	**Algebra** ✓ Graph data and find the gradient of a line. ✓ Solve exponential equations.
Activity 7 Allocating resources to maximize profit	✓ Revise understanding based on new information and evidence.	**Algebra** ✓ Use a linear programming model to help analyse a situation and make business decisions.
OTHER RELATED CONCEPTS	Equivalence Justification Model Pattern Representation	

Introducing simplification

Simplification is a key skill that permeates much of the mathematics that students learn. Starting with the order of operations, students need to learn the rules that allow them to move from a more complex form to a simpler one. This, in turn, allows them to solve a wide range of equations through application of the rules that they master.

However, simplification also appears in many other topics in mathematics. As a problem-solving strategy, "solving a simpler problem" is sometimes the logical place for students to begin. This is further developed in Pólya's problem-solving strategy, in Chapter 7 on Generalization.

Sometimes developing formulas beforehand can eliminate the need to perform the same type of work during similar, subsequent problems. Even finding a function to represent a graph can be simplified. This chapter allows students to work on a wide range of activities, each one putting their simplification skills to the test.

TOPIC 1

Simplifying algebraic and numeric expressions

To many students, the concept of simplification makes them think of their work with polynomials or the order of operations. Both of these involve using established rules to make an expression less complicated. In this topic students will be using both the laws of logarithms and the laws of exponents to achieve the same type of result. However, these laws will be applied to contexts that may surprise many students.

 Activity 1 **The geometric mean**

Whether students have been to a museum or simply looked at a poster on a wall, they can relate to the idea of finding the "best spot" in which to view something hanging in front of them. The task of finding the spot where the viewing angle is maximized is the theme of an old problem called the "Regiomontanus' angle maximization problem". It has a geometric solution that depends on the properties of angles inscribed in circles.

Before beginning the activity, display a poster in the classroom, and ask students to try to find the place that they think maximizes the viewing angle. Then they can test their response against the geometric mean of the distances to the top and bottom of the poster from the viewer's eyes.

Before they can simplify the formula for the geometric mean, students will need to know the laws of logarithms. It may be useful for students to try to develop them, based on a variety of examples, before deriving the laws algebraically.

TIP

Students and teachers may frequently refer to both the mean and average of data sets, using them interchangeably. Generally, they use the term "arithmetic mean" less often, although it is more descriptive of the value to which they are referring. Students know how to calculate it but they rarely know why it is called the arithmetic mean. Once they have learned about arithmetic sequences or progressions, explaining the reason for using the term in full can help solidify their understanding of it.

Try the teaching idea below, instructing students to round every value to the nearest thousandth.

TEACHING IDEA 1: The laws of logarithms

INQUIRY

a) Use a calculator to find these logarithms. Write the answers in the spaces provided.

i) $\log 2 =$ _____ $\log 3 =$ _____ $\log 6 =$ _____

ii) $\log 4 =$ _____ $\log 5 =$ _____ $\log 20 =$ _____

iii) $\log 2 =$ _____ $\log 10 =$ _____ $\log 20 =$ _____

iv) $\log 7 =$ _____ $\log 8 =$ _____ $\log 56 =$ _____

b) What do you notice about the answers in each row?

c) Give three different ways of calculating log 24 without actually entering "log 24" in the calculator.

d) Write a rule for what you discovered. Use appropriate mathematical notation and symbols.

e) How could you calculate log 5 if you knew log 30 and log 6? Write a rule similar to the one in question d) for division.

Students who struggle with this activity may need a hint to find the pattern among quantities in the same part of **a)**. For example, in part **(i)**, they should note that 2 and 3 multiplied together give 6. They will use this idea in question **e)**. They may need to be prompted to look at how the values they calculated in the same row relate to one another as well as how the operands of the logarithms relate to one another.

The viewing-angle problem can also be found in other areas. For example, a soccer player running down one of the sidelines of the field with the ball will want to kick it from the point where the angle to the goalposts is as large as possible. The NCTM website has an activity that allows students to adjust the position of the kicker in order to see the effect on the shooting angle. Simply search for "soccer problem" on the NCTM Illuminations website.

WEB LINKS

Visit http://illuminations.nctm.org and search for "Soccer Problem".

Activity 2   Dimensional analysis

Assessment

This activity is appropriate to use as a summative task. If you choose to assess students on this task, you can use criterion A. The task-specific descriptor in the top band (7–8) for criterion A should read that the student is able to:

- select the most appropriate exponent law to use when simplifying the formulas
- successfully apply these laws of exponents to simplify units in all of the given formulas
- solve these formulas correctly in a variety of contexts.

Stage 1 of the unit planner

Key concept	Related concepts	Global context
Form	Simplification Equivalence Justification	Scientific and technical innovation
Statement of inquiry		
Laws that govern objects on our planet can be justified, in part, by simplifying to produce equivalent forms.		

INTERDISCIPLINARY LINKS

This could be an opportunity for interdisciplinary work between sciences and mathematics. Students could be asked to perform dimensional analysis with units from the formulas that they will see in sciences throughout the year.

Dimensional analysis is sometimes called the factor label method. It often involves converting from one unit to another. In physics, it is also used to verify that the units are correct, when using a formula. Many physics teachers will encourage students to carry the units throughout their calculations and simplify them at the same time as performing the necessary mathematical operations on the quantities. Others prefer to let students perform this dimensional analysis first, allowing them to focus on the mathematics after the units have been verified. Despite the obvious connection to the laws of exponents, many students do not make the connection between the two.

For students to complete Activity 2 successfully, they will need to be well acquainted with the laws of exponents. You could use this teaching idea as a lead in.

TEACHING IDEA 2: The laws of exponents

All of the laws of exponents can be taught through an inquiry approach. Giving students plenty of practice in expanding expressions such as $x^2 \cdot x^5 = x \cdot x \cdot x \cdot x \cdot x \cdot x \cdot x = x^7$ allows them to discover, on their own, how the laws are formulated. Once these have been established, students can then apply these same laws to dimensional analysis.

 **DP LINKS**

Both exponents and logarithms are topics studied by students in DP mathematics at SL and HL.

Rather than simply showing students negative exponents, let them discover what they mean, using the activity below. Simply make sure that all students are writing their answers as whole numbers or fractions. Students who need more support may be guided to follow the pattern down any column. (For example, in the first column, each of the values is divided by two in order to obtain the one below it.) It is important that students discover the rule on their own and frame it in appropriate mathematical notation and symbols. When finished, be sure to discuss with students the scope, validity and limitations of their conjectures. Do they hold for rational exponents? How are their conclusions supported by other laws of exponents that they have learned? As seen in Chapter 7 on Generalization, conjectures must be subjected to rigorous proof in order to be validated.

TEACHING IDEA 3: Exponents

INQUIRY Change each expression to a whole number or fraction (no decimals). Write the answers in the spaces provided. Do not use a calculator. Simply follow the pattern.

a) $2^5 = \underline{\quad 32 \quad}$

$2^4 = \underline{\quad 16 \quad}$

$2^3 = \underline{\qquad}$

$2^2 = \underline{\qquad}$

$2^1 = \underline{\qquad}$

$2^0 = \underline{\qquad}$

$2^{-1} = \underline{\qquad}$

$2^{-2} = \underline{\qquad}$

$2^{-3} = \underline{\qquad}$

$2^{-4} = \underline{\qquad}$

$2^{-5} = \underline{\qquad}$

b) $3^4 = \underline{\qquad}$

$3^3 = \underline{\qquad}$

$3^2 = \underline{\qquad}$

$3^1 = \underline{\qquad}$

$3^0 = \underline{\qquad}$

$3^{-1} = \underline{\qquad}$

$3^{-2} = \underline{\qquad}$

$3^{-3} = \underline{\qquad}$

$3^{-4} = \underline{\qquad}$

c) $4^3 = \underline{\qquad}$

$4^2 = \underline{\qquad}$

$4^1 = \underline{\qquad}$

$4^0 = \underline{\qquad}$

$4^{-1} = \underline{\qquad}$

$4^{-2} = \underline{\qquad}$

$4^{-3} = \underline{\qquad}$

$4^{-4} = \underline{\qquad}$

d) Based on your results above, what conclusions can you draw about positive, negative and zero exponents?

e) Write rules for each of your conclusions. Use the correct mathematical symbols and notation.

INTERNATIONAL MATHEMATICS

In many regions of the world, exponents are also called "indices".

TEACHING IDEA 4: The laws of exponents—a different approach

INQUIRY   You could use this activity either to teach or to review the laws of exponents. Give students examples in which the laws of exponents are correctly applied and others in which they are not. From these, they discern the differences and arrive at the laws of exponents. This is also an effective form of review for students who have already been exposed to the laws of exponents.

Yes—these follow the rules	No—these do not follow the rules
$a^3 \times a^5 = a^8$	$a^4 \times a^3 = a^{12}$
$b^6 \times b^{10} = b^{16}$	$a^4 \times b^3 = ab^7$
$\dfrac{a^{10}}{a^2} = a^8$	$\dfrac{a^{10}}{a^2} = a^5$
$\dfrac{b^{26}}{b^{20}} = b^6$	$\dfrac{a^{10}}{b^2} = \left(\dfrac{a}{b}\right)^8$
$(a^4)^3 = a^{12}$	$(a^4)^3 = a^7$
$(b^6)^5 = b^{30}$	$(a^4)^3 = 3a^4$
$a^0 = 1$	$a^0 = 0$
$b^0 = 1$	$b^0 = b$

Ask students to summarize the laws of exponents. Make sure that they use correct mathematical symbols and notation. Use the next exercise to verify their understanding.

For each expression, determine whether or not the laws of exponents were applied correctly. If so, explain what rule was used. If not, give the correct answer, explaining the mistake and what should have been done.

Expression	Correct: What rule was applied correctly?	Incorrect: What is the correct answer? What mistake was made?
$m^6 \times m^5 = a^{30}$		
$(d^8)^2 = d^{10}$		
$\dfrac{y^{12}}{y^3} = y^9$		
$\dfrac{(2^4)^5}{(2^3)^2} = 2^3$		

Ask students similar questions with both positive and negative exponents to consolidate their understanding.

TEACHING IDEA 5: Units

Many formulas involve one or more constants, such as the gravitational constant G. Rather than giving students the units for the constant in a formula, have them deduce its units based on what units should result.

For example, the electric field inside a nuclear shell is calculated using $E = \dfrac{kqr}{R^3}$. Knowing that the electric field is measured in N/C and that q is measured in C, while both r and R are measured in metres, have students determine the units of k.

In the end, students can actually make up their own formulas and try to stump each other to either perform dimensional analysis correctly or determine the units of a constant.

TIP

In Activity 2, it is important for students to realize that they are working with units, not with the quantities themselves. The sum of two or more quantities that are measured in square metres produces another quantity, also measured in square metres. So, when a formula requires them to add m² and m², it does not result in 2m² but rather in m².

EXTENSION

If a film studio wanted to create a stunt in which the leaning tower of Pisa collapsed, the stunt producers (and students) could use dimensional analysis to work out how slow the camera should be to make the effect look real (assuming they use a model that is about 1 m tall).

Ask students what factors are likely to affect the time it takes for the tower to fall.

Their answers will probably include "the height of the tower", "the mass of the tower" and the "acceleration due to gravity". This could lead to a relationship such as:

$$t_{fall} = h^a \, m^b \, g^c$$

where the exponents a, b and c are to be determined (since we don't know how each quantity affects the time for the tower to fall). Assume that time (t_{fall}) is measured in seconds (s), and height h, in metres, mass m, in kilograms and g in m/s². Ask students to determine the exponents a, b and c so that the proper units result.

Solution

$$s = m^a \, (kg)^b \left(\frac{m}{s^2}\right)^c$$

Since the only unit that remains on the left is s, then:

i) $b = 0$ (the mass of the tower doesn't matter!)
ii) $a + c = 0$ (there are no metres on the left-hand side of the equation)
iii) $-2c = 1$

Therefore, $c = -\dfrac{1}{2}$ and $a = \dfrac{1}{2}$. The original formula becomes

$$t_{fall} = h^{\frac{1}{2}} g^{-\frac{1}{2}} \text{ or } t_{fall} = \sqrt{\frac{h}{g}}.$$

This describes the falling time of both the model and the actual tower of Pisa (if it fell!). If the height of the Leaning Tower of Pisa is about 100 m and the model measures 1 m (and the value of g is approximately 10 m/s² for both), it can then be shown that the camera should be slowed down to $\dfrac{1}{10}$ of normal speed for the fall of the model to look realistic.

TOPIC 2 ## Simplifying through formulas

Some mathematics students simply want to know the formula to solve problems. However, as explained in the student book, knowing a formula should never be the end in itself. Students should be able to develop the formulas that they will use and then be taught when to use them appropriately. So while the student book leads them to discover formulas for very specific problems, students should still be responsible for understanding their derivation and correct application.

👤 Activity 3 — Metric relations in a right-angled triangle

At the start of the activity (questions **a)** and **b)**), students are asked to use their maths knowledge to find the missing values. Typical solutions involve the use of similar triangles, Pythagoras' theorem and/or trigonometry. In the next part, students focus on similar triangles, from which they can derive the four common metric relations. Students who struggle may require more guidance. Let them work through these instructions.

a) Copy this right-angled triangle into your notebook.

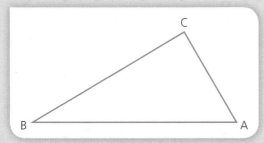

b) Label the sides appropriately.
c) Draw the altitude from right angle C to side AB. Label it h.
d) You should now have three triangles. Explain how they are related. Justify your answer.

Now label each of the segments of the hypotenuse, like this.

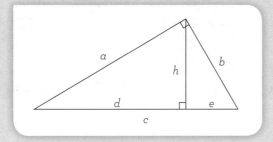

e) Based on your answer to **d)**, you should be able to write proportion statements relating sides of two triangles. Study the triangles and write proportion statements that relate each set of lengths

  (i) a, b, c and h **(ii)** d, e and h **(iii)** a, c and d **(iv)** b, c and e

f) Rewrite each of the proportions so that there is no division, only multiplication. These are the "metric relations".

TEACHING IDEA 6
Ask students to construct a right-angled triangle and include the altitude from the right angle to the hypotenuse. Prompt them to measure the various sides/segments and test the metric relations in their own triangles.

Activity 4 — Deriving formulas for other problems involving right-angled triangles

Students are asked to derive a formula that can be used to help solve problems involving double right-angled triangles, and hence make it easier to deal with similar questions in the future.

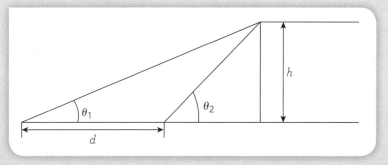

If students work with the appropriate right-angled triangles in these questions, they will use tangent ratios and will get the formula:

$$h = \frac{d\left(\tan\theta_1 \tan\theta_2\right)}{\tan\theta_2 - \tan\theta_1}$$

If students use the sine rule to work through the process, they will get the formula:

$$h = \frac{d\left(\sin\theta_1 \sin\theta_2\right)}{\sin\left(\theta_2 - \theta_1\right)}$$

Either formula will work.

TEACHING IDEA 7: How does the formula change?

Ask students to think about how the formula will change when the question is altered slightly.

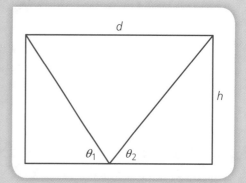

Trigonometry questions based on double right-angled triangles also come in a slightly different configuration.

Ask students to work through the process of deriving the formula again so that they notice the small difference.

Students who are using double right-angled triangles will not have many issues with the change. Students who use the sine rule may have difficulty seeing that $\sin(180° - \theta_1 - \theta_1) = \sin(\theta_1 + \theta_2)$ if they are not familiar with the unit circle. You may have to structure that section of the inquiry for them.

When they have derived the formula, students can try the following problem.

The photograph is of the Petronas Towers in Kuala Lumpur. They are identical in height and are the tallest twin towers in the world. There is a sky bridge on the 42nd floor that is 58.4 metres in length (to the centre of each of the two towers) and is positioned 170 metres above the ground. A tourist is standing on this bridge and can see the top of one tower, looking at an angle of elevation of 81.3°, and the top of the other tower, at an angle of elevation of 86.9°. Calculate the height of the Petronas Towers to the nearest metre.

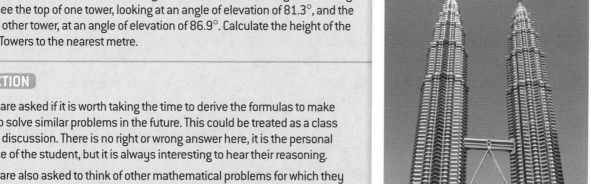

REFLECTION

Students are asked if it is worth taking the time to derive the formulas to make it easier to solve similar problems in the future. This could be treated as a class debate or discussion. There is no right or wrong answer here, it is the personal preference of the student, but it is always interesting to hear their reasoning.

Students are also asked to think of other mathematical problems for which they could try to derive a formula by using variables instead of numbers. You could ask them to work through the year's topics and see if they can make a list. They could do this on their own or in groups. You could assign different units to each group and gather a class set of formulas.

EXTENSION Deriving the quadratic formula

Using the standard form of the equation $ax^2 + bx + c = 0$, isolate x by completing the square.

What famous formula do you get?

Ask the students to research the history of the quadratic formula and who the first person was to derive it.

CHAPTER LINKS

Students may use the same methods and formulas from this topic in Chapter 9 on Measurement, where they are "measuring the immeasurable". The metric relations in Activity 3 could be used as a third method, provided that students find a position where the angle of elevation and the angle of depression total 90°. The double right-angled triangle formula derived in Activity 4 can be used directly in the "measuring the immeasurable" task, since students are using similar types of measurement in the second required method.

A good problem-solving strategy involves solving a simpler problem before tackling a more complicated one. There are times when you can simplify a problem by dividing it into smaller, more manageable parts. In this topic, students will learn how to do this with a variety of different mathematical content.

 Activity 5 **Polynomial equations**

STEP 1 **Quadratic equations**

Solving quadratic equations is a fundamental skill in algebra. However, many students still forget that neither the quadratic formula nor factorizing is appropriate if the equation is not written in the form $ax^2 + bx + c = 0$. It is definitely worth the effort to reinforce why this is so necessary. Ask students questions such as:

"If $AB = 12$, what do you know for sure about the value of A or B, other than their signs?"

"What if $AB = -8$? What do you know for sure about the value of A or B?"

You could either then elicit what value AB would have to take, in order for there to be some guarantee, or simply ask what they would know for sure if $AB = 0$. Understanding this will also allow them to attempt to solve higher degree polynomial equations.

STEP 2 **Other polynomial equations**

Students will need to know how to perform long division on polynomials before beginning this task.

In this part of the activity, students are asked to divide a polynomial by a binomial (most probably by using long division). They are expected to reason that $(x + 1)$ is **not** a factor of $x^3 - 2x^2 - 5x + 6$ because the division produces a remainder. However, $x - 1$ **is** a factor since the remainder is zero. Once they have this factor, they know that $x^3 - 2x^2 - 5x + 6 = (x - 1)(x^2 - x - 6)$. From here they should be able to factorize the quadratic factor to give a factorized form of $(x - 1)(x - 3)(x + 2)$.

In the remainder of the exercises, students will need to guess and check potential factors. Students who struggle may need some hints as to what factors to try. The first equation is factorizable so no division is actually necessary. The last two have $(x + 2)$ as a factor. Students could be given a hint such as: "Try $(x + 1)$, $(x - 1)$, $(x + 2)$ or $(x - 2)$." Once they have found a factor, they should be able to divide and factorize the remaining quadratic. They can then solve each equation.

EXTENSION

Factorizing higher degree polynomials is a skill that will allow students to solve polynomial equations and graph polynomial functions. Students who can find factors, based on simple trial and error, may be introduced to the remainder theorem, the rational zeros theorem and even synthetic division. All of these will provide students with a systematic method for finding factors, thereby simplifying the entire process.

Students could also use dynamic geometric software (DGS) to explore the shapes of graphs.

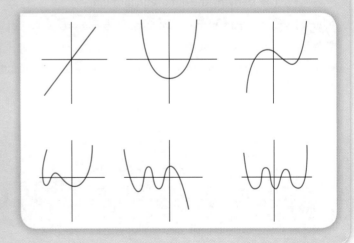

Activity 6 Dating the Earth

Key concept	Related concepts	Global context
Relationships	Simplification Pattern Model	Scientific and technical innovation
Statement of inquiry		
Establishing and modelling relationships between variables can help us understand the origins of our planet.		

Assessment

This activity is appropriate to use as a summative task. If you choose to assess students on this task, you can use criterion D. The task-specific descriptor in the top band (7–8) for Criterion D should read that the student is able to:

- graph the given data correctly
- find an equation to represent the data
- interpret the gradient correctly and use it to determine the age of the Earth
- justify the degree of accuracy for their estimate of the age of the Earth
- justify whether their estimate of the age of the Earth makes sense.

In this activity, students are introduced to what may happen when there are three variables and they want to draw a graph in two dimensions. Because P and D change as time (t) passes, there is no way to represent the three variables in a two-dimensional coordinate system. However, if just P and D are graphed, then the slope of the graph should be equal to $e^{\lambda t} - 1$. In this way, the age of the meteorites and the Earth (t) can be determined from the gradient of the line.

In the reflection at the end of the activity, students are asked what to graph in order to simplify the process of finding the equation relating F and r. Since the relationship between them is an inverse squared relationship, then students should graph F against $\frac{1}{r^2}$.

Particularly ambitious students might like to research the use of parametric equations.

TIP

When students draw the graph of D against P, they will need to create a line of best fit, using the method(s) learned in class. When they set the gradient equal to $e^{\lambda t} - 1$, they will then need to use logarithms to determine the value of t. This makes this activity a useful review of a variety of skills and concepts.

Systems of inequalities

Linear programming can be used to help analyse situations and lead to better business decisions. In this activity, students write a set of linear inequalities to represent a real-life scenario and then graph the system. Since the linear programming model represents a small company that sells two products, students will analyse the points of intersection (POI) in order to determine how the company can maximize its profitability.

TEACHING IDEA 8: Moving from graphing an equality to graphing an inequality

INQUIRY Ask students to complete this exercise.

a) State what the inequality $y \leq 3x - 9$ means, in words.

b) What is the difference between an equality $(y = 3x - 9)$ and an inequality $(y \leq 3x - 9)$?

c) Test these points to see if they satisfy the equality and the inequality.

Does it satisfy the relation?	$y = 2x - 4$	$y \leq 2x - 4$
Test Point A $(5, 6)$		
Test Point B $(8, -4)$		
Test Point C $(-7, 2)$		
Test Point D $(-1, -10)$		

d) Find three more test points that satisfy one relation but not the other. How many test points are there that satisfy the inequality but not the equality?

e) State in words the difference between the graphs of an equality and that of an inequality. How would you graph an inequality?

f) Graph the inequality on a set of coordinate axes. How is it related to the graph of the equality?

g) What happens when the inequality condition is just "greater than" or "less than", as opposed to "greater than or equal to" or "less than or equal to"?

Choose three test points, including a test point that would fall on the line $y = -\frac{1}{2}x + 3$ to see if the line is included in your inequality.

h) Graph these two inequalities to consolidate your understanding, selecting only one test point.

$$y < -4x + 6 \qquad y \geq \frac{2}{3}x - 1$$

TIP

Students may find it easier to determine the feasibility region in a linear programming model if they shade the section that is not included in the inequality, because the feasibility region will stay clear with no shading in it. It does not matter how many constraints you have to graph in your model, the section required will always be free of shading and it may be easier to determine the vertices. This is especially true if students are going to use their GDC to graph the inequalities.

Graphing an Inequality as a Jigsaw Activity

Before class, prepare four groups of materials/activities and place each group in a different corner of the room.

Group 1
Materials where students review/learn and practise rewriting a linear equation into gradient–intercept form $(y = mx + b)$.

Group 2
Materials where students review/learn and practise how to graph a line in gradient–intercept form.

Group 3
Materials where students learn and practise how to shade the graph of an inequality.

Group 4
Materials where students learn and practise the difference between the graph with "<" or ">" and one with "≤" or "≥".

Arrange students in groups of four and number the students in each group 1–4. Ask each student to go to the corner of the room that has the same number as they have been allocated within their group. There, in a new group, they must learn or review the content that is there. (Remind them that they are responsible for taking this knowledge back to their home group, since every student within that group is learning something different.) When they have completed their activities in the four different corners, they return to their home group and complete, as a group, exercises in which they graph inequalities, starting with Step 1. One of the benefits of this activity is that you can supervise the grouping and match the students to the "corner" with the appropriate difficulty level to challenge them.

Assessment

This activity is appropriate to use as a summative task at the end of a unit on systems of inequalities. If you choose to assess students on this task, you can use criterion D. The task-specific descriptor in the top band (7–8) should read that the student is able to:

- determine all inequalities, constraint functions and the original objective function correctly
- graph the original scenario correctly
- determine the optimal solution of the original scenario
- use technology to solve all three scenarios correctly
- justify why their optimal solution makes sense, including the effect of constraints
- explain clearly the relationship between the algebraic and graphical representation of profit and how it validates their solution
- justify all assumptions made in alternate scenarios
- use qualitative data to effectively make and justify decisions.

Stage 1 of the unit planner

Key concept	Related concepts	Global context
Relationships	Simplification Model Representation	Scientific and technical innovation
Statement of inquiry		
Better business decisions can be made by modelling relationships to create a simplified representation of real-life scenarios.		

TIP

If the students are setting up the equations and using software such as Excel to help them, then they will be able to have more than two variables and many constraints. If you would like them to graph their model by hand then it's best to keep it manageable, and limit them to two variables so it can be done on the two-dimensional Cartesian plane.

STEP 2 **Adapting the linear programming model**

This task demonstrates that a company is never stagnant because it has to be able to adapt to the changing market (and other factors) that influence its profitability. Once the model is built, using Excel or other spreadsheet software, a company would then alter the numbers or constraints to analyse how it can best adapt to these changes. To reflect this, students are asked to repeat Task 2, using graphing software.

⊂⊃ CHAPTER LINKS

In Chapter 6 on Equivalence the students use a system of equations to plan a balanced diet. This could be adapted so that they use the same ideas to create a linear programming model.

EXTENSION

Once students have a good understanding of linear programming, you could encourage them to set up their own linear programming model. For example, they could try developing the most nutrient-rich diet for developing countries, based on their two staple foods. Ideally, the students will suggest their own ideas. This would be far more creative and could then be a summative task.

Summary

In mathematics, students often think "simplification" refers to reducing expressions to a less complicated form. However, developing formulae to solve complex problems is another aspect of this important concept. Because formulas are not always possible or desirable, solving a simpler problem may help find the solution to a more complex one. In this chapter, students have explored these different facets of "simplification".

Space

The frame of geometrical dimensions describing an entity

	ATL skills	Mathematics skills
TOPIC 1	Special points and lines in 2D shapes	
Activity 1 Get to the point!	✓ Listen actively to other perspectives and ideas. ✓ Propose and evaluate a variety of solutions.	**Algebra** ✓ Solve systems of linear equations. **Geometry and trigonometry** ✓ Recognize lines of symmetry of shapes. ✓ Recognize the properties of polygons. ✓ Calculate the coordinates of midpoints and the distance between points. ✓ Use properties of triangles to find the location of their special points (centres).
TOPIC 2	Mathematics and art	
Activity 2 Properties of 3D objects and impossible objects	✓ Develop new skills, techniques and strategies for effective learning.	**Geometry and trigonometry** ✓ Explore the characteristics of the platonic solids. ✓ Analyse plane sections of solids.
Activity 3 The Koch snowflake	✓ Develop new skills, techniques and strategies for effective learning.	**Algebra** ✓ Solve problems involving geometric sequences and series. **Geometry and trigonometry** ✓ Calculate perimeters and areas of triangles.
TOPIC 3	Volume and surface area of 3D shapes	
Activity 4 Let's pack it in!	✓ Create novel solutions to authentic problems.	**Number** ✓ Calculate the percentage increase and decrease of surface area and volume to determine the best configuration. **Geometry and trigonometry** ✓ Calculate the surface area and volume of rectangular prisms and cylinders. ✓ Calculate perimeters and areas of plane figures. ✓ Use Pythagoras' theorem to find unknown side lengths.
OTHER RELATED CONCEPTS	Quantity Model	Representation

Introducing space

In this chapter students will first explore Cartesian methods to represent points and lines, calculate distances and determine positions of points of intersection of lines. They will be invited to learn more about Rene Descartes, a major figure in the 17th century, who made invaluable contributions to both sciences and mathematics.

The research activity below puts into context the terminology, notation and general concepts that students will be learning and using in the chapter activities. However, it is not necessary to have completed this before doing the chapter activities.

> We must admit with humility that, while number is purely a product of our minds, space has a reality outside our minds.
>
> Carl Friedrich Gauss

TEACHING IDEA 1: Rene Descartes and coordinate geometry

The goal of this research is to discover how Descartes' work changed the way mathematicians looked at geometrical problems. In particular, students can be invited to answer these questions.

a) What is a Cartesian coordinate system?

b) How can it be used to represent points, lines and other geometric objects?

c) Why is Descartes said to have married algebra and geometry?

d) How has the work of Descartes led to the discovery of the area of mathematics called calculus, and to the overall progress of science?

e) Are there other coordinate systems? What purpose do they serve?

WEB LINKS

Go to www.storyofmathematics.com and search for Descartes.

Search online for Cartesian coordinate system; www.cut-the-knot.org has a good example.

Go to www.algebra.com and search for different types of coordinate systems.

TOPIC 1 Special points and lines in 2D shapes

Students need to recognize symmetrical regular shapes as special, with particular properties. For example, the lines of symmetry meet at a point called the centroid.

Finding the centroid of non-regular shapes, such as non-equilateral triangles, is not easy. Students will need to use their knowledge of other special lines in a triangle, such as medians, angle bisectors, altitudes and perpendicular bisectors of sides and, if possible, should be guided to explore the relation between these lines for different types of triangle.

Activity 1 Get to the point!

Before attempting this activity, students should be familiar with coordinate geometry and know:

- how to use Cartesian coordinates to represent points
- the relation between coordinates of the midpoint of a segment and coordinates of its extreme points (endpoints)
- how to find the equation of a line, given two of its points
- how to determine the point of intersection of two lines.

Asking students to make mobiles is an excellent way for them to discover the importance of special lines and points. They will be able to assess their success while constructing the mobile, as well as in the finished product. Additionally, they can have fun with the use of colour and different shapes!

Students should recognize that the lines of symmetry meet at a point called the centroid. For non-regular shapes, such as a scalene triangle, identifying the centroid will require knowledge of other special lines in a triangle, such as medians, angle bisectors, altitudes and perpendicular bisectors of sides. Students may need guidance to explore the relation between theses lines for the different types of triangle.

These two teaching ideas will help students see the properties of special points and lines more clearly, since they are not dependent on any handmade inaccurate measurements. They can explore further, using DGS, after the hands-on approach the activity requires.

TEACHING IDEA 2: Exploring triangle properties

As a start, you could introduce students to dynamic geometry software (DGS). This will help them understand the definitions and concepts for the activity in the student book, and facilitate observing relationships and patterns.

Let students use DGS to draw a triangle and then the special lines:

- medians—lines defined by each vertex and the midpoint of the opposite side
- altitudes—lines through each vertex and perpendicular to the opposite side
- perpendicular bisectors of the sides—lines through the midpoint of each side perpendicular to that side
- angle bisectors—rays that bisect the angles.

They should label the vertices of a triangle A, B and C, then drag these points to obtain different types of triangles and investigate under which conditions some of these lines overlap (for example, for equilateral triangles each median is also an altitude, a perpendicular bisector and an angle bisector).

Alternatively, if DGS is not available, students may use paper models of different triangles to explore the properties listed above (a cut-and-fold activity).

As they complete this task, they may explore other patterns: each set of lines meets at a point. This is a good opportunity for students to research special points of a triangle, sometimes called centres of a triangle: the barycentre where the medians meet, the orthocentre where the altitudes meet, the circumcentre where the perpendicular bisectors meet and the incentre where the angle bisectors meet.

◌◌ INTERNATIONAL MATHEMATICS

The barycentre may be referred to as the centroid depending on where you are in the world.

TEACHING IDEA 3: Exploring special points in a triangle

DGS is a wonderful tool for further exploration of the special points of a triangle, and to help students justify the conjectures they make, based on patterns they have observed.

Ask your students to use dynamic geometry software for these tasks.

- Draw a triangle PQR and find its circumcentre C. Draw a circle with centre C and radius PC. Drag the points P, Q and/or R and write down your conjecture. Research and find an application of the circumcircle of a triangle in the real world.
- Draw a triangle PQR and find its incentre I. Draw a circle with centre I and a tangent to the side PQ. Drag the points P, Q and/or R and write down your conjecture. Research and find an application of the incircle of a triangle in the real world.

◌◌ DP LINKS

This leads into the geometry topic of further mathematics.

- Draw a triangle PQR and find its orthocentre H, the circumcentre C and barycentre G.
- Draw the line CH and GH. Drag the points P, Q and/or R and write down your conjecture.
- Research and learn more about the Euler line and the nine-point circle.

STEP 3 Students should recognize the usefulness of Pythagoras' theorem in its application to finding the distance between two points.

Further exploration

After students have completed the activity, it might be interesting to go further and explore an alternative, simple way to determine the location of the barycentre of a triangle when the coordinates of its vertices are known.

Ask your students to use dynamic geometry software to draw triangle PQR and find its barycentre G. Drag the vertices of the triangle a few times and complete a table like this one.

$P(x_1, y_1)$	$Q(x_2, y_2)$	$R(x_3, y_3)$	$G(x_G, y_G)$	$x_1 + x_2 + x_3$	$y_1 + y_2 + y_3$

They should then deduce a formula for the centroid of the triangle in terms of the coordinates of its vertices.

◯◯ DP LINKS

In DP mathematics, HL students are sometimes asked to use equations of planes in 3D to determine the position of the barycentre of a tetrahedron. This activity offers a sleek alternative method to solve this problem.

EXTENSION

Check that the result holds in 3D and use it to determine the centroid of a tetrahedron, using barycentric coordinates.

TOPIC 2   Mathematics and art

Mathematics has always been a central feature of many artistic forms in every culture and in every century, both pre-historic and historic— from early cave drawings, Egyptian hieroglyphics and pyramids, Greek architecture, symmetrical patterns in mosques, Renaissance sculpture and painting, all the way to modern art. Not only in visual arts, but also in music, dance, photography, and other art forms, mathematics is ever present.

 Activity 2   **Properties of 3D objects and impossible objects**

This activity attempts to highlight some of the ways that mathematics is incorporated into works of art, both 2D and 3D.

STEPS 1–3 Students will focus on the well-known platonic solid, the cube, and explore its properties. They should establish some relations between 2D and 3D objects, and recognize 2D shapes as sections of 3D objects.

In step 3 **a)**, students are advised to use potatoes for this activity. Instead of using potatoes, however, you might want to let your students use DGS to explore the sections of a cube.

This table shows the results that students should aim for. They may need gentle hints from time to time!

Example	Polygon	Slicing plane
	Equilateral triangle	It is produced by any plane perpendicular to one of the diagonals of the cube that intersects three of its faces.
	Isosceles triangle	It is produced by a plane that isn't perpendicular to any of the diagonals of the cube, that intersects three faces of the cube and that is parallel to the diagonal of one of these faces.
	Other triangles	It is produced by any other plane that intersects three faces of the cube.

The section produced by any plane that intersects two pairs of parallel faces of the cube is a parallelogram.

Example	Polygon	Slicing plane
	Square	It is produced when a cube is sliced by a plane parallel to one of its faces.
	Rectangle	It is produced when the cube is sliced by a plane parallel to one of its edges that intersects four of its faces. It may be a square if the edges of the rectangles are congruent.
	Trapezium	It is produced when the slicing plane intersects four faces of the cube and just two of these faces are parallel. If the slicing plane is parallel to the diagonals of the parallel faces then the trapezium is isosceles.
	Pentagon	It is produced by a plane that intersects five of the six faces of the cube. It is impossible to obtain a regular pentagon because the pentagon produced has two parallel sides.
	Hexagon	It is produced when the slicing plane intersects all the faces of the cube. This is a regular hexagon if the slicing plane is perpendicular to one of the diagonals of the cube and contains the midpoints of the edges of the cube that it cuts.

It is impossible to produce a polygon with more than six sides!

Here are some possible extensions to the activity.

TEACHING IDEA 4: Platonic solids

What makes platonic solids special? Invite students to do a research project on platonic solids, and create a display about their properties. They can then attempt to deduce a formula for the area and volume of the solid in terms of the length of an edge.

WEB LINKS
For descriptions of solids see www.3quarks.com.

TEACHING IDEA 5: 3D to 2D

Research the work of the artist Albrecht Dürer and the technique he uses to draw sections of a cone. Use DGS to illustrate this method and explain how he went about moving from 3D to 2D without distorting the shape of the section.

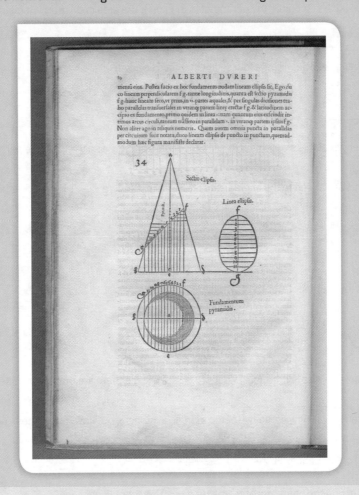

Fractal geometry

Although fractal geometry is not yet part of most standard middle- and high-school curricula, this emerging area of mathematics has already redefined concepts such as measurement, form and dimension. You can use this exciting new area of mathematics to revise and consolidate some standard geometrical ideas and concepts, such as geometric sequences and series. Students worldwide appreciate the intrinsic beauty of fractal forms. They are fascinated to discover that their most-loved film animation characters and landscapes are generated, to a large extent, through fractal geometry. In this sense, a fractal can be considered mathematically as algorithmic art!

Activity 3 The Koch snowflake

As well as introducing students to the properties of fractals, the Koch snowflake provides unusual and interesting results, such as an object with an infinite perimeter having finite area. Students should also appreciate the aesthetic designs produced in generating fractals.

Students might need help in correctly applying the instructions. You might want to describe what happens to one side of the original equilateral triangle as the recursion is repeated.

a)

STEP 0 To start with, one side of the equilateral triangle, or one of its segments, has a length of 1.

Since there are three such segments, the perimeter is $3(1)$ or 3. The area of the equilateral triangle is $\frac{1}{3}$.

STEP 1 One side of the equilateral triangle will have four segments, each of length $\frac{1}{3}$.

Therefore the total length of this side is now $4\left(\frac{1}{3}\right)$. Since there are three such sides, the perimeter is $3(4)\left(\frac{1}{3}\right)$ or 4.

Since three segments are added to the original, the area is increased by $3\left(\frac{1}{9}\right)$. The area is therefore $\frac{\sqrt{3}}{4} + \frac{\sqrt{3}}{4}\left(\frac{3}{9}\right)$

or $A = \frac{\sqrt{3}}{4}\left(1 + \frac{3}{9}\right) = \left(\frac{1}{\sqrt{3}}\right)$.

STEP 2 There are four times four segments, each of length $\left(\frac{1}{3}\right)\left(\frac{1}{3}\right)$.

The total length of this one side is therefore $\left(\frac{4}{3}\right)\left(\frac{4}{3}\right)$ or $\left(\frac{4}{3}\right)^2$. Since there are three such sides, the perimeter is

$3\left(\frac{4}{3}\right)^2$, or $\frac{16}{3}$. The triangle will now contain $\frac{1}{9}$ of the triangle in stage 1, but now four of these per side are being added

on, for a total of $3(4)$ or 12. Hence: $A = \frac{\sqrt{3}}{4}\left(1 + \frac{3}{9} + 4 \cdot \frac{3}{9^2}\right) = \left(\frac{10}{9\sqrt{3}}\right)$

Students are asked to find the perimeter and area for $n = 3$. They should get:

$P = 3\left(\frac{4}{3}\right)^3 = \frac{16}{3}$ and $A = \frac{\sqrt{3}}{4}\left(1 + \frac{3}{9} + 4 \cdot \frac{3}{9^2} + 4^2 \cdot \frac{3}{9^3}\right) = \left(\frac{94}{81\sqrt{3}}\right)$ before considering the nth stage.

They should now be able to see that, at stage n, $P = 3\left(\frac{4}{3}\right)^n$ and, with some help,

$A = \frac{\sqrt{3}}{4}\left(1 + \frac{3(4)^{n-1}}{9^n}\right)$.

---

b) Students should now consider the quotients of the expressions for one stage and the previous stage, to observe the relationship of the perimeter and area of stage n to the previous stage $(n-1)$. This is the relationship between successive stages. In this way, both P_n and A_n can be described as geometric progressions. Additionally, A_n is an infinite geometric series such that $|r| < 1$ and hence its sum can be found by applying the formula.

c) The formulas are already partially generated above in considering one segment of the original triangle at each successive stage.

d) Students will have observed that, as n increases, the perimeter increases rapidly and the area decreases rapidly—the infinite perimeter of the Koch snowflake is contained in a finite area!

EXTENSION 1

a) Explore whether it is possible to tessellate a plane with copies of the Koch snowflake that are the same size, of two different sizes, or of more than two different sizes.

b) Write a computer program to create the Koch snowflake.

c) Research and create a 3D Koch snowflake.

d) Research these ideas.

- Fractals formed by modifying the Koch snowflake
- L-systems fractals
- Fractal dimension
- Real-world applications of fractals

⊂⊃ WEB LINKS

Visit www.shodor.org. Go to **Activities and lessons>Interactive>Activities,** then search for Koch's snowflake.

Also visit eurekaelearning.com. Go to Upper Secondary and search for Sierpinski carpet.

EXTENSION 2 **Slices of a 3D fractal**

Students have explored the slices of cubes within a cube and can now explore slices of a famous fractal cube, the Menger sponge.

The Menger sponge is formed by removing a cube within the cube in an iterative manner. With every iteration, its area approaches infinity and its volume approaches zero!

Taking slices of the Menger sponge produces interesting figures, in particular, diagonal slices. Students may be asked to guess what kinds of shape will be produced during the slicing sequences, and to check their answers on the website listed here.

⊂⊃ WEB LINKS

To view the slices, visit www.simonsfoundation.org, then search for "Menger sponge slice".

TOPIC 3 Volumes and surface area of 3D shapes

This task shows students that the mathematical analysis of a real-life situation can be used to help determine optimal solutions, such as the most efficient allocation of resources. Ultimately, it will make them reflect upon their role as a consumer when looking at the packaging of products.

Assessment

This activity is appropriate to use as a summative task. If you choose to assess students on this task, you can use criterion D. The task-specific descriptor in the top band (7–8) for criterion D should read that the student is able to:

- create 2D diagrams of <u>all</u> possible configurations that are easy to follow and can be used effectively to help visualize and solve the task
- accurately calculate the surface area of all possible configurations
- determine the optimal configuration to minimize the surface area and wasted space in the box and use the least amount of cardboard possible
- accurately calculate the wasted space in the box for the optimal configuration
- accurately calculate the percentage change of the surface area and wasted space, comparing the original configuration to the optimal configuration
- justify the degree of accuracy used in calculations, given the context of the task
- critically reflect on the optimal configuration, given the context of the problem and its suitability in real life, and justify your recommendation on which packaging strategy the company should use.

WEB LINKS

A website with all of the configurations, and a built-in calculator where you simply have to input the number of circles, is available to help you visualize the problem in 2D and give formulas to find the area and perimeter (not recommended to be shared with the students, at least not initially!). Visit www.had2know.com. (Click on **Calculators>Mathematics>Geometry>** Hexagonal Circle Packing)

Stage 1 of the unit planner

Key concept	Related concepts	Global context
Form	Space Measurement	Globalization and sustainability
Statement of inquiry		
Designing optimal configurations for packaging three-dimensional forms in space allows us to use resources more responsibly.		

Before attempting this task, students should have already learned the relevant skills.

Inquiry activities on how to teach some of these skills are:

- deriving the volume formulas of 3D shapes by taking the face shape (2D) and multiplying by the third dimension—this can also be done as a review if they were taught the formulas in a previous year
- deriving the surface area formulas of 3D shapes by breaking each shape down into its 2D net—the use of manipulatives really helps students with this as they create the 3D shape then simply break it apart, lie it flat and break it down into 2D shapes.

Problem-solving activities on how to teach some of these skills include:

- finding the volume that is wasted when one or multiple shapes are packed into another—you could ask students to find an example of one shape packaged inside another and tell them to measure all necessary dimensions. With a class set of examples, students can then determine the wasted space and amount of material required to make the packaging of as many of these examples as you wish. Alternatively, you could have a sports theme and find the wasted volume and surface area packaging for a pack of three tennis balls or squash balls or a sleeve of golf balls.

TEACHING IDEA 6: Minimizing surface area for a set volume

INQUIRY Give students a set volume and ask them to determine what 3D shape will minimize surface area. This task leads well into the idea of packaging and efficient use of resources, and gives excellent practice in transposing and solving algebraic equations. As students discover that it is a sphere, you can have reflection questions about the practicality of using a sphere versus other 3D shapes. You can guide them through this by telling them which 3D shapes to look at, or leave it open for them to decide.

Summative task: Optimal packaging

This activity leads students to recognize that there is an optimal configuration for cylindrical cans in a rectangular box that will minimize wasted space and surface area packaging.

Encourage students to suggest as many different configurations as possible—you may want to tell them to think beyond the regular traditional packing configurations. You could even provide empty cans and let the students move them around, to try to discover novel configurations. You may choose to tell them to work in 2D, as the height of the soup can will not affect the configuration choice. Ultimately, you would like them to suggest a hexagonal arrangement.

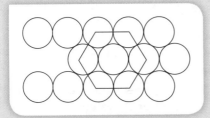

Having diagrams really helps with the calculations, especially if they have discovered the hexagonal arrangement, because they will need to use Pythagoras' theorem to determine the dimensions.

TIP

This task is excellent as a summative assessment task because all students should experience success and see it through to completion, even if they do not have the optimal configuration. As an extension of this task, you could ask students to derive the formulas to find the area and perimeter of the 2D rectangle that borders the circles.

EXTENSION 3D structures

Students could research how mathematics has influenced the work of architect Richard Buckminster Fuller. They should write a report describing how he applied his philosophy of "more for less" when he designed buildings and structures such as the geodesic dome in Montreal.

Geodesic dome in Montreal

Summary

Defining points in space, as Descartes did was the beginning of a journey into the world of geometry. Over the last 400 years, mathematicians have developed more geometric tools, enabling them better to define our 3D universe and analyse more efficient ways of managing limited earthly resources. The last 50 years have witnessed the birth of a new geometry, which is newly defining dynamic phenomena and the concept of dimension.

CHAPTER

16 System

A group of interrelated elements

	ATL skills	Mathematics skills
TOPIC 1 The real number system		
Activity 1 No shortcuts allowed!	✓ Use and interpret a range of discipline-specific terms and symbols.	**Number** ✓ Explain the components and properties of the real number system. **Algebra** ✓ Solve equations algebraically.
Activity 2 Testing the axioms	✓ Use and interpret a range of discipline-specific terms and symbols.	**Number** ✓ Explain the components and properties of the real number system.
Activity 3 Is being "irrational" like being "complex"?	✓ Use and interpret a range of discipline-specific terms and symbols.	**Number** ✓ Explain the components and properties of the real number system. ✓ Transform between different forms of number. ✓ Simplify numerical expressions. **Algebra** ✓ Expand and simplify algebraic expressions.
TOPIC 2 Geometric systems		
Activity 4 Creating an axiomatic system	✓ Understand and use mathematical notation and terminology.	**Geometry and trigonometry** ✓ Explore the components and properties of a geometric system.
TOPIC 3 Probability systems		
Activity 5 Using axioms	✓ Understand and use mathematical notation.	**Algebra** ✓ Use set notation and terminology. ✓ Determine the intersection and union of sets. **Statistics and probability** ✓ Prove basic probability theorems. ✓ Apply the properties of probability systems to solve problems.
Activity 6 Blood typing 1	✓ Make connections between subject groups and disciplines.	**Statistics and probability** ✓ Apply the properties of probability systems to solve problems.
Activity 7 Blood typing 2	✓ Understand and use mathematical notation.	**Statistics and probability** ✓ Apply the properties of probability systems to solve problems.
Activity 8 Bringing a ship to dock	✓ Demonstrate persistence and perseverance.	**Statistics and probability** ✓ Prove basic probability theorems. ✓ Apply the properties of probability systems to solve problems.

Introducing systems

The notion of a mathematical system, while easy to define, is one of the most abstract of mathematical constructions. Students largely learn properties of number systems and Euclidean geometry during the MYP years. This chapter begins with the basic building blocks of any axiomatic system, and will especially serve those students who go on to study more advanced mathematics after the MYP, where a good grounding in the basic elements of pure mathematics is advantageous.

> *In principle, we can know all of mathematics. It is given to us in its entirety and does not change... That part of it of which we have a perfect view seems beautiful, suggesting harmony; that is that all the parts fit together although we see fragments of them only.*
>
> Kurt F Gödel

TOPIC 1   The real number system

This topic is intended to introduce students to the properties of real numbers under the binary operations of addition and multiplication.

A binary operation on a set is a calculation that combines two elements of the set.

Students have been working with numerous kinds of binary operation, without identifying them as such. As well as the arithmetic operations with which students are familiar, the set operations of union and intersection are also binary operations. Before starting the activities in this topic, students should know the definitions of the properties: commutative, associative, identity, inverse, and distributive.

TIP

Binary operation tables

A visual way to test the properties of binary operations is to use a table like the one here. The binary operation ♥ on the set {m,a,t,h} gives the results in the table. Students can write out the results of the operations from the table. Some questions to pose are:

a) How can we tell from the table whether or not the set under the operation has an identity?

b) Using the table, determine if each element has an inverse, and if so, write them down.

♥	m	a	t	h
m	h	m	a	t
a	m	a	t	h
t	a	t	h	m
h	t	h	m	a

 ### Activity 1   No shortcuts allowed!

If students know the properties of the real number system, with the binary operations of addition and multiplication, then they should be able to use them as justification for the steps in solving an equation. Alternatively, they can work through the first part of the activity, research the properties and give an example of each. Then, they use these definitions to determine whether the properties hold under subtraction and division.

It is important to go through part **c)** with them to make sure that they understand the justification for each stage in solving an equation. When they know how to use the properties to solve an equation, they should then attempt to use them in part **d)** to justify the solution process of the other equations. They conclude with defining subtraction and division in terms of addition and multiplication, respectively.

Activity 2   Testing the axioms

In this activity, students will determine whether certain properties hold in the given sets under the given binary operations. It is fairly straightforward to test for the commutative and associative properties.

Although they are not asked to test for the other properties, you could ask them to do so as extension exercises.

Students often confuse the identity and inverse properties, so it might be beneficial to go over a few of these with them. They must test for the identity first, since the inverse is defined in terms of the identity. They must also test for the right and left identities and inverses, since both must be commutative.

Activity 3 | **Is being "irrational" like being "complex"?**

In this activity, students will perform operations on irrational numbers and complex numbers, and note similarities in the operations.

TEACHING IDEA 1: Polynomials and expressions with square roots

Before beginning step 2, let students practise adding, subtracting and multiplying polynomials. (You might want them to justify each stage, as in Activity 1.) Then let them begin the exercises in step 2 without giving detailed instructions. Most students will be able to work out how to answer the questions without help, although they may need some guidance in how to multiply irrational numbers of the form $a\sqrt{b}$.

TEACHING IDEA 2: Rationalizing denominators

Inevitably, some students will ask why they need to rationalize denominators at all, or why they are not asked to rationalize numerators. The main reason for rationalizing a denominator is to change it into a more useful form. This simplifies the task of performing certain arithmetic operations. Before computers and calculators were readily available, students were taught how to find the square root of a number. Certainly, if you know how to take the square root of a number, it is easier to find, for example, $\frac{\sqrt{2}}{2}$ than it is to find $\frac{1}{\sqrt{2}}$.

Recognizing similarities between math topics is an excellent way of reviewing content. It also helps students understand current techniques in light of what they already know. When demonstrating how to rationalize a denominator, you could start by asking students to practise multiplying binomials such as $(x-4)(x+4)$ in order to see why the process of rationalizing the denominator works.

Emphasize that, in rationalizing the denominator, students are really multiplying the fraction by an expression that is equivalent to 1.

Once students are comfortable with performing operations on irrational numbers, including rationalizing the denominator, they are ready for similar operations with complex numbers. They may need reminding that, with complex numbers $i^2 = -1$, which often allows them to simplify expressions further.

> **TIP**
>
> Some students make the connection between operations with complex numbers and simplifying algebraic expressions, since i looks like a variable. Ensure they also make connections to operations with irrational numbers since i represents $\sqrt{-1}$, a radical.

TOPIC 2   Geometric systems

⊘ CHAPTER LINKS

Chapter 8 on Justification has more on direct and indirect proofs.

This topic is intended to familiarize students with elements of an axiomatic system. These include undefined terms, defined terms, axioms or postulates, and theorems. It would be worthwhile reviewing with them the geometry they already know in terms of these elements. In Euclidean geometry, "point" is the only undefined term. The definitions of lines, planes and so on follow from there. It is important to impress upon the students that they may only accept the smallest possible set of axioms, or statements without proof, and then use these to prove theorems. Showing them a few examples of Euclidean geometry, for example, proving that the sum of the angles of a triangle is 180°, should impress upon them two of the properties of an axiomatic system: it is concise and consistent.

Students should be ready to use finite geometry to construct their own axiomatic system in this activity. As a preparation for this activity, you might want to set this task.

Instructions to students: Three-point geometry
Undefined terms: **point** and **line**

Axioms:

A1: There exist exactly three distinct points.

A2: Any two distinct points are on exactly one line.

A3: Not all the points are on the same line.

A4: Any two distinct lines pass through at least one common point.

What does this geometry look like? Try to make a picture.

Using the axioms, determine how many lines this three-point geometry has. Prove your conjecture.

Suggested proof
A1, A2 and A3 together imply that there are at least three lines. Suppose now there is a fourth line. This line must then share a point with the other lines. Therefore, this line must connect two already connected points. Since each two distinct points are on exactly one line, no fourth line can exist. QED.

Once students have completed this task, it seems logical to ask what a four-point geometry could look like. Students might like to devise their own axioms. They should reach this conclusion.

Undefined terms: **point** and **line**

Axioms:

A1: There are exactly four points.

A2: Every pair of distinct points is on exactly one line.

A3: Each line has exactly two points on it.

Make a possible drawing of this four-point geometry.

Some possible drawings are shown below.

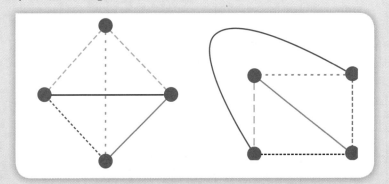

Make a conjecture about how many lines are in a four-point geometry, and prove your conjecture.

TIP

Students should realize that lines need not be straight, and may come up with a variety of representations, such as the examples below.

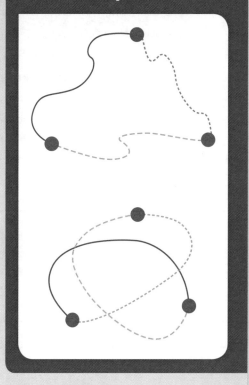

TIP

Suggested proof: Use A2 and A3 to count how many pairs of points there are, label the four points from A1 as P, Q, R and S. Making all possible lines from these points, you have PQ, PR, PS, QR, QS and RS, that is, six lines.

Now, prove that each point of the four-point geometry has exactly three lines on it.

Students are now ready to tackle the activity in the student book.

Here are some sample responses to question **c)**.

TIP

Suggested proof: A2 implies that each point has a line in common with the other points. Hence, by A1, there are at least three lines on every point. Now suppose that there are four lines on a point. Then, by A3, this fourth line must be on some other point. However, by A2, any two points are on exactly one line. Hence, there are exactly three lines on every point.

(i) A1 implies that a line exists and A2 implies that the line contains exactly three points. Call this line l_1 and label the points A, B and C.

(ii) A1 and A2 imply that at least three points exist. A3 implies that at least one more point, D, exists that is not on l_1. Hence, A1, A2 and A3 imply that at least four points exist.

(iii) A1 and A2 imply that there is a line and it contains three points. A3 implies the existence of a point not on that line.

It might be helpful to draw a table, such as the one below.

l_1	l_2					
A	D					
B						
C						

A4 implies the existence of lines containing A and D, B and D, and C and D.

l_1	l_2	l_3	l_4			
A	D	D	D			
B	A	B	C			
C						

A2 implies that each line must have three points, and A4 implies that these must be new points.

l_1	l_2	l_3	l_4			
A	D	D	D			
B	A	B	C			
C	E	F	G			

A4 implies that each pairing of E with one of the other points determines a line. Since E is already paired with A and D on l_2, this leaves B, C, F and G. Since B and F and C and G have already been paired off, this situation creates two new lines.

l_1	l_2	l_3	l_4	l_5	l_6	
A	D	D	D	E	E	
B	A	B	C	B	C	
C	E	F	G	G	F	

Using the same argument, F has not been paired with either A or G. This necessitates one more line. G is now paired with every other point, and since every point is now paired with every other point, A5 is satisfied.

l_1	l_2	l_3	l_4	l_5	l_6	l_7
A	D	D	D	E	E	F
B	A	B	C	B	C	G
C	E	F	G	G	F	A

Any attempt to create more lines will go against an axiom. Hence, this geometric system has seven lines and a minimum of seven points. But, since each line contains three points already, there cannot be more than seven points.

(iv) There are many models that can correctly depict this geometric system. Students need to be reminded that lines need not be straight. Arcs and segments will need to be incorporated in their model.

TEACHING IDEA 3: Applying an axiomatic system

Students can use the axiomatic system they have just created to solve the following problem.

INSTRUCTIONS TO STUDENTS

Your school has designated you to form a sporting club consisting of teams that will represent the school at the annual international schools' sporting tournament. The rules are as follows.

R1: At least one team will be formed.

R2: Every team will have exactly three runners in it.

R3: Not all runners are on the same team.

R4: For every pair of runners, there is exactly one team that contains the pair.

R5: Every two teams have at least one runner in common.

a) Draw a diagram that shows your club's requirements for entry into the tournament.

b) Determine the minimum number of runners that can be entered into the tournament.

c) Prove that your answer to part b) is also the maximum number of runners.

d) Find the minimum number of teams you can enter into the event.

e) Prove: Every two distinct teams have exactly one runner in common.

The solutions for a)–d) are essentially the same as those of the activity in the student book. The teams are the lines and runners are the points. Students can use the same diagram they used in c)(iv).

For part e), using the given information, there are two distinct teams l_1 and l_2. By R5, these teams share one or more runners. Hence, these two teams are the same team! This is false, since these teams are distinct. Now, suppose these teams share two runners. By R4, any two runners determine exactly one team. Again, this is false because the two teams are distinct. Now suppose these teams have no runners in common. This contradicts R5, since every two teams must share one runner. Hence, since all other possibilities are false, the two teams must share one common runner.

TOPIC 3 Probability systems

It is fairly easy to prove well-known probability results from their systems of axioms. It is important to make clear to students that there is really no difference between a theorem, a proposition and a corollary when it comes to proving them. The distinction in terms of terminology comes from the importance of the results. Usually a theorem is an important result while a proposition is less important. A corollary is something that follows easily from a theorem or proposition that has just been proved (for example, a particular case).

In the student book, students are given an example of a system of axioms for probability and asked to deduce well-known probability theorems. It is a good idea to support the proofs with Venn diagrams. Most of the results require use of disjoint sets to allow the application of axiom A3. These disjoints sets can be obtained by splitting the sets. Again, Venn diagrams help in visualizing convenient ways of splitting the sets. The following activity is an example of how these theorems can be proved.

👤 Activity 5 ╲ Using axioms

In this activity, students will use the axioms A1–A3 and any theorem already proved in the topic introduction to deduce the results.

Answers

T2: $P(A') = 1 - P(A)$, for any $A \subseteq S$.

Proof:

By definition of the complementary set, $A \cup A' = S$ and $A \cap A' = \varnothing$.

Then by axiom A1, $P(A \cup A') = P(S)$.

By axiom A3, as $A \cap A' = \varnothing$, $P(A \cup A') = P(A) + P(A')$.

Therefore, $P(A) + P(A') = 1 \Rightarrow P(A') = 1 - P(A)$. QED.

T3: If $S = \{a_1, \ldots, a_n\}$ is a sample space where all the single outcomes have the same probability, then for any outcome A,

so that $n(A) = k$, $p(A) = \dfrac{k}{n}$.

Proof:

Consider the single outcomes $A_1 = \{a_1\}$, $A_2 = \{a_2\}, \ldots, A_n = \{a_n\}$.

Let $p = P(A_1) = \cdots = P(A_n)$.

As the single events are disjoint,

$$P(A_1 \cup A_2 \cup \cdots \cup A_n) = P(A_1) + P(A_2) + \cdots + P(A_n) = p + p + \cdots + p = np$$

As, $A_1 \cup A_2 \cup \ldots \cup A_n = S$, $P(A_1 \cup A_2 \cup \ldots \cup A_n) = P(S) = 1$.

Therefore $np = 1 \Rightarrow p = \dfrac{1}{n}$ and, if $A = A_{x_1} \cup A_{x_2} \cup \ldots \cup A_{x_k}, x_1, x_2, \ldots, x_n$ are distinct values from the set 1, 2, …, n.

$$P(A) = P\left(A_{x_1} \cup A_{x_2} \cup \cdots \cup A_{x_k}\right) = P\left(A_{x_1}\right) + P\left(A_{x_2}\right) + \cdots + P\left(A_{x_k}\right) = \underbrace{\frac{1}{n} + \frac{1}{n} + \cdots + \frac{1}{n}}_{k \text{ times}} = \frac{k}{n} \quad \text{QED.}$$

T4: $P(A \cup B) = P(A) + P(B) - P(A \cap B)$, for any events A and B.

Proof:

First notice that $A \cup B = A \cup (B \cap A')$.

As A and $B \cap A'$ are disjoint sets (here a Venn diagram helps), by A3,

$$P(A \cup B) = P(A \cup (B \cap A')) = P(A) + P(B \cap A')$$

Then notice that B can be written as the union of two disjoint events:

$B = (B \cap A) \cup (B \cap A')$ and therefore

$P(B) = P((B \cap A) \cup (B \cap A')) = P(B \cap A) + P(B \cap A')$

$\therefore P(B \cap A') = P(B) - P(B \cap A)$ (2)

By (1) and (2), $P(A \cup B) = P(A) + P(B) - P(B \cap A)$. QED.

T5: $P(\varnothing) = 0$

Proof:

$S \cup \varnothing = S$ and $S \cap \varnothing = \varnothing$.

By theorem 2, $P(\varnothing) = 1 - P(S)$ and, by axiom A1, $P(S) = 1$.

Therefore, $P(\varnothing) = 1 - P(S) = 1 - 1 = 0$. QED.

Blood typing

When students have read the introduction to the task in the student book, and completed the first activity, direct them to the award-winning blood typing game found at www.nobelprize.org. (go to educational>Play the Blood Typing Game). The website explains that the purpose of this educational game is for players to learn the basics about human blood types and blood typing, as well as to understand why it is important—to enable medical practitioners to save lives by performing safe blood transfusions. Another purpose is to offer a game experience that is challenging and fun, of course!

Students will be challenged to save patients in need of blood transfusions. They will have to determine the blood type of the patient, and then decide on the best compatibility for the patient, from the blood that is available.

 Activities 6 and 7 **Bloody typing 1 and Blood typing 2**

These blood typing activities are intended to encourage students to use the probability axioms in a real-life setting.

INTERDISCIPLINARY LINKS The activity can be used alone, or as an interdisciplinary activity. It could be connected with biology as part of a "cell" unit, looking at cell behaviour and immunological reactions, or in a "genetics" unit, studying genetically determined blood group systems (of which A, B, O and Rh blood systems are considered in this task as they are used for blood transfusions) and genetic blood disorders such as haemophilia. See the teaching idea for some possible research questions.

Assessment

These activities are appropriate to use as summative tasks. If you choose to assess students on these tasks, you can use criterion D. The task-specific descriptor in the top band (7–8) for criterion D should read that the student is able to:

- identify the real-life elements of the blood typing game
- select appropriate mathematical strategies to accurately categorize the blood types both as donors and recipients
- apply the mathematical strategies to reach correct solutions to the various real-world situations posed
- justify the degree of accuracy the mathematical strategies selected to model the real-life problem posed.

Key concept	Related concepts	Global context
Relationships	Systems Representation	Scientific and technical innovation
Statement of inquiry		
Using a system to represent and analyse relationships has led to biological advances that have benefited mankind.		

Answers

The answers for the chart in Activity 6 are as follows.

Blood type	Can donate blood to	Can receive blood from
O+	O+, A+, B+, AB+	O+, O−
O−	All types (universal donor)	O−
A+	A+, AB+	A+, A−, O+, O−
A−	A−, A+, AB−, AB+	A−, O−
B+	B+, AB+	B+, B−, O+, O−
B−	B−, B+, AB−, AB+	B−, O−
AB+	AB+	All types (universal recipient)
AB−	AB−, AB+	AB−, A−, B−, O−

The answers for the chart in Activity 7 are as follows.

Blood type	Can donate blood to	Can receive blood from	% of general population	Probability of finding a compatible donor
O+	O+, A+, B+, AB+	O+, O−	37.4	44%
O−	All types (universal donor)	O−	6.6	6.6%
A+	A+, AB+	A+, A−, O+, O−	35.7	86%
A−	A−, A+, AB−, AB+	A−, O−	6.3	12.9%
B+	B+, AB+	B+, B−, O+, O−	8.5	54%
B−	B−, B+, AB−, AB+	B−, O−	1.5	8.1%
AB+	AB+	All types (universal recipient)	3.4	100%
AB−	AB−, AB+	AB−, A−, B−, O−	0.6	15%

a) Students should justify using A3 since the events are mutually exclusive, that is, a person cannot have two different blood types.

b) (i) and (iv).

c) Students will need to use conditional probability to calculate the answers, for example, $P(A|Finland) = \dfrac{0.22}{0.5} = 0.44$. This does not equal $P(A)$.

EXTENSION 1

Now that students have some knowledge about blood types, they could be encouraged to do some research to enhance their understanding. The following could serve as a guide as to which questions to research.

- How is a person's blood type determined?
- Does blood type vary according to gender, race and ethnicity, or geographic location?
- How do your parents' blood types influence yours?
  What possible blood types can an offspring have, depending on the blood types of the parents?
- What happens if a person receives an incorrect blood type?
  An incorrect rhesus factor?
- How much blood can you lose before you need a blood transfusion?
- Is there any danger in donating blood to a blood bank?
  How many times a year is it recommended that anyone should donate blood?
- Should we donate blood on a regular basis?
  Why, or why not?

EXTENSION 2

Group activity

Create numbered cards for your class, with blood type and rhesus factor on them in the same proportion as the statistics for blood types for the US that are given in the student book. Alternatively, you can use statistics from the country that you live in.

Suppose you have a class of 20 students. According to the chart in the student book, about 37% of the US population has blood of type 0+, so you would need to assign about 37% of 20 cards to this blood type; this would be about seven cards. (Alternatively, you can use figures from the country you live in to make it more realistic.)

Continue in this way for all the blood types. The students are identified by the number on top of the card, rather than their own name.

My blood type	Compatible types				Non-compatible types			
Label each column as M (match), C (compatible) or N (non-compatible).								
	0+	0−	A+	A−	B+	B−	AB+	AB−
M, C, N								

Each student should randomly select a blood-type card from those you have prepared. Then they write the blood type stated on the card at the top of their own card, and fill in the compatible and non-compatible blood types.

They then fill in each column with:

- M for a match
- C for compatible
- N for non-compatible.

You will also need a master chart to record the class results.

The blood type game – experimental probability of survival
Determining your experimental probability of survival

Label each box as Y (survives) or N (does not survive).

M, C, N										
n-trial number	0+	0−	A+	A−	B+	B−	AB+	AB−	P(S) as fraction	P(S) as decimal
1										
2										
3										
4										
5										
6										
7										
8										
9										
10										
11										
12										
13										
14										
15										
16										
17										
18										
19										
20										

Now, students imagine that there has been a disaster. The victims (who are represented by students) are all rushed to the hospital. (You can change the scenario if you feel this would affect your students.) Due to low blood supplies, there has been a call for donors.

Create 100 cards representing the different blood types in the same proportion as the general population. Shuffle the cards so that the students can select them randomly.

Each student takes a turn to present themselves as a victim arriving at the hospital. They pick up the next card, from the pile of 100 blood-type cards that represent the donors. (If you prefer, the students could be the donors and the blood-type cards could represent the victims.)

The students estimate their probability of survival, based on the donor's blood type and the details on their own card. Students record the blood type, and whether that blood type is a match, compatible or non-compatible to the students' selected blood type.

The probability of survival will therefore be $P(S) = \dfrac{P[M]}{n}$ or $P(S) = \dfrac{P[C]}{n}$, where n is the trial number.

Alternatively, $P(S) = 1 - \dfrac{P[N]}{n}$ The students add these on to each successive trial, and divide by the trial number.

After they have recorded the probabilities, students can graph their results on a grid, with the trial number on the horizontal axis and $P(S)$, as a decimal, on the vertical axis. The graph represents their experimental probability of survival.

To represent their theoretical probability of survival, students use the chart indicating the percentage of occurrence of blood types in the general population. They add the percentages of a match together with compatible blood types, change the sum to a decimal and draw a horizontal line on their graph at this decimal value.

Discussion points can include how the theoretical and experimental probabilities differ, and how many trials are needed so that these probabilities would be the same.

Exploring properties of axiomatic systems

In this final activity, students will play a simple game, to reinforce their understanding of the 3C's of a system: consistent, complete and concise.

 Activity 8 **Bringing a ship to dock**

The ship to dock activity is to remind students of the three characteristics of an axiomatic system: it must be complete, concise and consistent.

There are many solutions to the puzzle. Two possible solutions are: SHIP, SLIP, SLOP, SLOT, SOOT, LOOT, LOOK, LOCK, DOCK or SHIP, SHOP, CHOP, COOP, COOK, DOOK, DOCK. Perhaps your students can manage even fewer moves!

Summary

System is one of the most abstract concepts in the study of mathematics. The intention is that, through the activities in both the student and teacher book, this concept has become clearer for students. It is very important that students realize that all of mathematics is based upon rigorous yet simple and elegant (and yes, even beautiful!) logic. The study of mathematics therefore can indeed help to improve the reasoning processes with which they tackle problems in their everyday lives.

Notes

Notes

Notes

Notes

Notes

Notes

Notes